# MECHANICAL IMPACT DYNAMICS

# MECHANICAL IMPACT DYNAMICS

## RIGID BODY COLLISIONS

**RAYMOND M. BRACH**
University of Notre Dame

A Wiley-Interscience Publication
**JOHN WILEY & SONS**
New York / Chichester / Brisbane / Toronto / Singapore

***Library of Congress Cataloging in Publication Data:***

Brach, Raymond M.
Mechanical impact dynamics / Raymond M. Brach.
p. cm.
"A Wiley-Interscience publication."
Includes bibliographical references and index.
ISBN 0-471-50845-4
1. Impact. I. Title.
QA937.B73 1991
531'.38—dc20 90-39860
CIP

Printed in the United States of America

10 9 8 7 6 5 4 3 2 1

To my wife, Joyce

# PREFACE

Low-speed collisions are ubiquitous—hammers on nails, hail on roof tops, athletic balls on floors and walls, screen doors, vehicles, typewriter carriages, dot matrix printers, shot peening, solid particles in multiphase flows, and so forth. The classical theory of the dynamics of colliding solid objects (and its conduit to reality, the coefficient of restitution) has an unusual place in the field of mechanics. It is covered in almost all high school physics courses. Most, if not all, engineering and science students learn it during their college studies. Yet the subject is then dropped and left undeveloped. Although the theory follows directly from Newton's laws of motion, it is said by some to be too simple and simplistic to be of any practical value. In some cases, the assumption of instantaneous velocity changes may not be appropriate. In others, the coefficient of restitution may depend significantly on the initial velocity. But all theories have limitations and, in many cases, classical impact theory provides useful results. More and more of these applications arise every day, and the subject deserves to be improved and expanded.

The theory is unusual for another reason. The mathematics associated with Newtonian and Lagrangian dynamics is intrinsically tied to differential equations, more often than not nonlinear differential equations. On the other hand, the equations of impulse and momentum are algebraic and almost always linear. It is this simplicity that makes this approach to impact a useful engineering tool. It is this classical approach that is covered in this book.

The primary goal of *Mechanical Impact Dynamics* is to develop and illustrate methods that allow the calculation of changes in rigid body motion, specifically velocity changes, due to an impact, and not to provide a compendium of well-known methods and results from mechanics. Topics such as the dynamics of falling chains and the center of percussion are already well covered in existing dynamics books. The classical book on impact by Goldsmith (see Ref. 8 in Chapter 1) still provides an excellent blend of theory and experiment and should always be consulted as a primary reference. The classical approach to impact dynamics followed in this book ultimately deals with algebraic equations, not differential equations. For this reason, almost the entire book can be understood by those who have completed a course in undergraduate dynamics. In particular, it should be useful to the practicing engineer who has an impact problem to deal with. The exceptions to the algebraic formulation occur in parts of Chapters 3 and 9, which deal with local deformation and oscillatory systems, respectively. By and large the methods and applications covered in this book are original and have been developed and used by the author over the last 15 years. The rationale for this book is that it provides a broader coverage of the classical theory. *Mechanical Impact Dynamics* presents a new approach to the treatment of friction or, more appropriately, tangential processes including indentation. The common assumption of "point contact" is eliminated and the concept of contact moment impulses is introduced. The recognition and treatment of tangential restitution are covered. Another new topic included here is that of collisions accompanied by permanent shape changes of one or both bodies. Improper handling of "friction" in collisions has often caused difficulties and has led to serious errors. In the experimental treatment of particles colliding with a flat surface, an entire field of literature contains claims that the friction coefficient depends on the angle of approach. In fact, the friction coefficient appears in some cases to be a constant; it is the impulse ratio that changes with angle of approach. A general treatment of tangential impulses is presented that exposes their important relationship to kinetic energy loss, permitting the proper representation of tangential effects.

A good amount of experimental data is presented from the field of erosion and wear due to particle impact. Applications to other fields are presented. An entire chapter is devoted to the impact of automotive vehicles. A method of modeling impacts of manipulators, or robotic arms, is described. The relationship of vibratory impact to the subject of chaos is covered. The velocity changes of a charged aerosol particle bouncing between diverging plates are included in Chapter 9. The applications covered here can only begin to touch the innumerable potential applications in engineering, meteorology, biomechanics, and other fields. For this reason, the coverage is kept to as fundamental a level as possible. Examples are presented to illustrate the theory, not to provide a catalog of applications.

One of the reasons for the simplicity of classical impact theory is that it uses coefficients, namely, coefficients of restitution and friction, to represent all of the nasty behavior that occurs at the interface in an impact. For some applications, estimates of coefficient values may be made analytically. In many situations, experimental values are necessary. These coefficients vary with material, body geometry, temperature, and initial velocities. As such, these coefficients are not constants (although they are treated as mathematical constants in equations). That these coefficients are not "constants" has led some to feel that classical impact theory has little utility. But if one has an impact dynamics problem to solve, he or she has only to look at the alternatives, such as an elastic–plastic finite element analysis; a few experiments may begin to look attractive. A more substantive shortcoming of classical theory is that it does not furnish forces and stresses inside of the colliding objects. It does provide velocity changes, impulses, and energy loss, which can sometimes be used (with force–time curve shapes) to estimate forces and stresses. Although it does not directly provide forces, displacements, and stresses, in many situations, such as high-speed vehicle collisions, classical theory has the remarkable ability to model a horribly complicated mechanical process. In the long run, it is the algebraic simplicity which will ensure that classical theory will be applied over and over again by engineers faced with impact problems.

Efforts such as preparing a book can never be done by one person. An author, including this one, needs the help of many others. John Kozol and Salvatore Ferrara, graduate students, and numerous undergraduates at Notre Dame have contributed to the development of individual topics. My son, R. Matthew Brach, and colleagues, Terry Akai, Mike Stanisic, and Mihir Sen, contributed immeasurably by their suggestions after reading portions of the text. Much appreciation is also due to Marilyn Walker, who turned scribbles into a manuscript.

RAYMOND M. BRACH

*Notre Dame, Indiana*
*November 1990*

# CONTENTS

# NOTATION

Note that throughout this book lower case letters are used to denote initial velocities, for example, $v$ and $\omega$. Capital letters are used to represent final velocities, such as $V$ and $\Omega$.

| | |
|---|---|
| $A$, $B$, $C$ | defined by Eqs. (5.40); coefficient matrices, Chapter 2 |
| $B_i$ | defined by Eq. (5.18) |
| $b$ | distance |
| $C$ | contact point; constraint impulse |
| $c$ | damping rate |
| $d$, $D$ | distance |
| $e$ | coefficient of normal restitution |
| $e_t$ | coefficient of tangential restitution |
| $e_m$ | coefficient of moment restitution |
| $F$ | force |
| $f$ | coefficient of Coulomb friction |
| $G$ | center of gravity |
| $g$ | acceleration of gravity |
| $H$ | angular momentum |
| $h$ | distance |
| $I$, $J$ | mass moment of inertia |
| $\bar{I}$ | $I_1 I_2/(I_1 + I_2)$ |

| | |
|---|---|
| $k$ | stiffness; radius of gyration |
| $l$ | length |
| $M$ | moment impulse |
| $m$ | mass |
| $\overline{m}$ | $m_1 m_2/(m_1 + m_2)$ |
| $N$ | moment of a force; number of bodies in a chain |
| $n$ | coordinate normal to a surface |
| $P$ | impulse |
| $Q$ | sum of squares of deviations (least square curve fitting) |
| $q$ | defined by Eq. (5.33) |
| $R$ | radius or distance |
| $R_p$ | pitch radius |
| $r$ | initial velocity ratio, tangential to normal; index |
| $T$ | kinetic energy |
| $T_L$ | kinetic energy loss |
| $t$ | coordinate tangential to a surface |
| $U$ | linear momentum variable |
| $V$ | velocity at the end of a collision |
| $v$ | velocity at the beginning of a collision |
| $v_{rn}$ | defined by Eq. (5.36) |
| $W$ | angular momentum variable; work |
| $x, y, z$ | coordinates |
| $\alpha$ | angle of approach; relative deformation, Chapter 3 |
| $\beta$ | angle; frequency ratio, Chapter 6 |
| $\Delta$ | defined by Eq. (5.58) |
| $\delta$ | defined by Eq. (5.17) |
| $\gamma$ | defined by Eq. (5.17) |
| $\Gamma$ | defined by Eq. (5.58); angular coordinate |
| $\zeta$ | viscous damping ratio |
| $\eta$ | angular coordinate |
| $\theta, \Theta$ | angular coordinate |
| $\lambda$ | dimensionless ratio, $R^2/k^2$ |
| $\lambda_M$ | impulse ratio, Eq. (6.39) |
| $\mu$ | ratio of the tangential impulse component to the normal impulse component |
| $\mu_0$ | impulse ratio corresponding to zero final relative tangential velocity at the contact area |
| $\mu_T$ | impulse ratio corresponding to zero energy loss |
| $\mu_m$ | impulse ratio that maximizes the kinetic energy loss |

| | |
|---|---|
| $\mu_c$ | critical impulse ratio that provides a limiting value of $\mu$ for use in solution equations |
| $\tau$ | time |
| $\phi$, $\Phi$ | angular coordinate |
| $\Omega$ | final angular velocity |
| $\omega$ | initial angular velocity; vibrational frequency |

# MECHANICAL IMPACT DYNAMICS

# CHAPTER 1

# PRINCIPLES, ASSUMPTIONS, AND DEFINITIONS

Most of the material in this book can be classified as applications of Newtonian mechanics to collision problems. A few of the fundamentals of Newton's second law and how they relate to the collision problem are discussed in this chapter. A simple example involving the momentum of a sliding mass is presented to illustrate a concept often treated incorrectly in impact analyses. Assumptions that usually are made when applying the principles of impulse and momentum to the impact process are discussed.

## 1.1 PRINCIPLES OF PARTICLE DYNAMICS

Newton's second law for a particle of mass $m$ is

$$F_r(\tau) = m\frac{dv_r}{d\tau}, \qquad r = x, y, z \tag{1.1}$$

where $r$ signifies a coordinate direction—$x$, $y$, or $z$. $F_r$ is the component in the $r$ direction of the resultant force acting on the particle and $\tau$ is time. The quantity $v_r(\tau)$ is the velocity component of mass $m$ in the $r$

direction. Equation 1.1 can be integrated over the path of motion from the point $(x_1, y_1, z_1)$ to the point $(x_2, y_2, z_2)$ to give

$$\tfrac{1}{2}mv_2^2 - \tfrac{1}{2}mv_1^2 = \int_1^2 (F_x\,dx + F_y\,dy + F_z\,dz) \tag{1.2}$$

or, in a more compact form,

$$\Delta T = W_{12}$$

These are expressions of the principle of work and energy over a displacement path from point 1 to point 2, where $W_{12}$ is the work done by the resultant external force and $\Delta T$ is the change in kinetic energy. If Eq. 1.1 is integrated with respect to time, the result can be written as

$$mv_r(\tau_2) - mv_r(\tau_1) = \int_{\tau_1}^{\tau_2} F_r(\tau)\,d\tau = P_r, \qquad r = x, y, z \tag{1.3}$$

The result is a vector where Eq. 1.3 is an expression of the $r$th component. $P_r$ is called the impulse component of the force component $F_r(\tau)$, and the product of the mass and the velocity is called the linear momentum component. Consequently, Eq. 1.3 shows that change in momentum is equal to the impulse over an arbitrary time interval.

The same forms of Eqs. 1.1, 1.2, and 1.3 are valid for rigid bodies as well as particles and provide the displacements, velocities, and accelerations of the mass center. A relationship similar to Eq. 1.3 holds for rotational motion. For rotational motion, the analog of Eq. 1.1 is

$$N_r(\tau) = \frac{dH_r}{dt}, \qquad r = x, y, z \tag{1.4}$$

where $H_r$ is the angular momentum component about the $r$ axis and $N_r$ is the corresponding component of the resultant moment. This equation can be integrated; the result is

$$H_r(\tau_2) - H_r(\tau_1) = \int_{\tau_1}^{\tau_2} N_r(\tau)\,d\tau = M_r, \qquad r = x, y, z \tag{1.5}$$

Here, it is seen that the change in angular momentum along the $r$ axis is equal to the moment impulse component written here as $M_r$. If $r$ lies along a principle axis of the rigid body, Eq. 1.5 can be written simply as

$$I_r\omega_r(\tau_2) - I_r\omega_r(\tau_1) = M_r \tag{1.6}$$

in which $I_r$ is the moment of inertia about the $r$ axis and $\omega_r$ is the corresponding component of the angular velocity; the product is the angular momentum.

Equations 1.3 and 1.6 are expressions of the impulse-momentum principle and form the basis of almost all of what follows in this book.

## ■ *Example 1.1*

PROBLEM: A particle of mass $m$ has an initial velocity $v$ at time $\tau_1$ and slides along a horizontal surface that has a constant coefficient of friction $f$. The particle has a velocity $V$ at time $\tau_2$ after sliding a distance $d$. Determine the

a. impulse, $P_f$, developed by friction in bringing the particle to rest, $V = 0$;
b. kinetic energy loss, $T_L$;
c. distance it slides to rest;
d. velocity of the particle if, after coming to rest, an additional negative impulse is applied equal to 10% of the frictional impulse found in part a;
e. ratio of energy lost in part d to that of part b.

*Solution:* From Eq. 1.3, the impulse $P_f$ to bring the particle to rest, $V = 0$, is

a. $P_f = mV - mv = -mv$

Since the particle is brought to rest, all of its kinetic energy is lost. From Eq. 1.2, this is

b. $T_L = -\Delta T = \frac{1}{2}mv^2$

Since the change in kinetic energy is equal to the work done

$$T_L = fmgd$$

This gives the sliding distance, $d$, where

c. $d = v^2/2fg$

For a total impulse of $1.1P_f$, the final velocity will be

$$V = v + \frac{1}{m}(1.1P_f)$$

d. $V = v + \frac{1}{m}(-1.1mv) = -0.1v$

This shows that the additional impulse reverses the direction of the particle's motion. By Eq. 1.2, the kinetic energy loss now becomes

$$T_{Ld} = \tfrac{1}{2}mv^2 - \tfrac{1}{2}mV^2 = 0.99\,(\tfrac{1}{2}mv^2)$$

e. $T_{Ld}/T_L = 0.99$

Note that the particle's energy loss now is less than before, since the additional impluse has done work on the particle, restoring 1% of its original kinetic energy. The additional impluse also reverses the direction of sliding. ■

## 1.2 SIMPLE PARTICLE IMPACT

### 1.2.1 Planar Collision of Two Smooth Particles

Numerous textbooks on dynamics, for example, Refs. 1–7, do a thorough job of presenting the fundamentals of dynamics as sketched in the preceding section. Most do not develop the treatment of collisions beyond a rudimentary level. One of the best is Pestel and Thomson,[6] but even their coverage stops short of a comprehensive treatment of the subject. In this section, simple impacts are covered at a level typically found in an introductory dynamics textbook. Figure 1.1 shows various categories of impacts. Coverage in this section is restricted to direct and oblique central, planar impacts of two particles. Frictional and other shearing effects are neglected.

Figure 1.2 shows free body diagrams of two particles and an associated, inertial coordinate system. For collisions, the classical problem is to relate the final velocities to the initial velocities. The following conditions are assumed to apply here:

1. Initial velocity vectors lie in the $n$–$t$ plane.
2. Angular velocities are negligible.
3. Deformation is small; contact occurs at a point on each mass.
4. The normal axis passes through each mass center and the point of contact.
5. Surfaces are smooth (frictionless); no tangential forces are generated at the contact point.

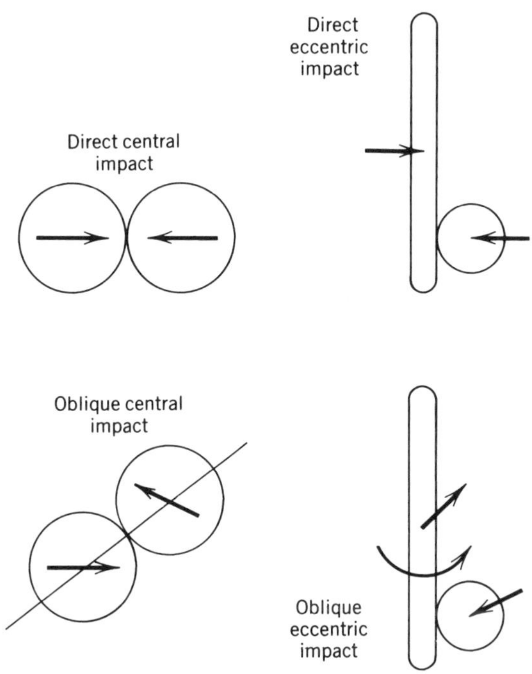


**Figure 1.1.** Basic collision categories.

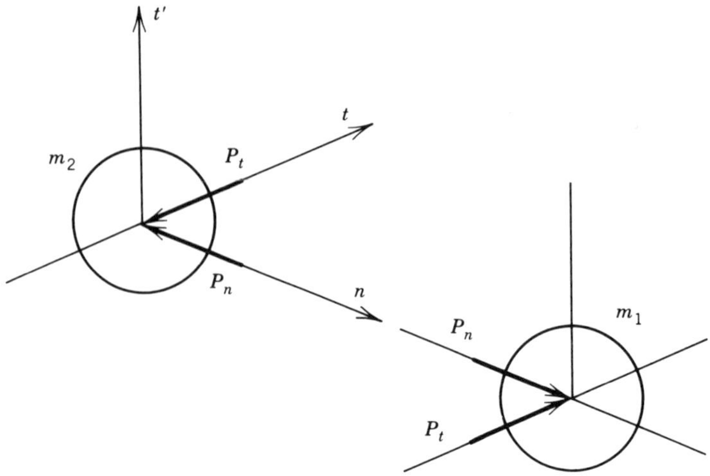


**Figure 1.2.** Free body diagram of colliding particles.

6. The duration of contact is short.
7. The normal force due to the impact of the two masses is large enough such that all forces except the normal force generated by the impact can be neglected.
8. During the duration of contact, the displacements are infinitesimal, velocity changes are finite, and accelerations are infinite.

Each mass has two initial velocity components. These components are $v_{1n}$, $v_{1t}$ and $v_{2n}$, $v_{2t}$ for a total of four. The same is true for final velocity components, $V_{1n}$, $V_{1t}$ and $V_{2n}$, $V_{2t}$. If the initial velocities are known and the final velocities unknown, four equations are required for a solution. Equation 1.3 can be applied to each mass; this gives two equations for $r = n$ and two equations for $r = t$. These are

$$m_1 V_{1n} - m_1 v_{1n} = P_n \tag{1.7}$$

$$m_1 V_{1t} - m_1 v_{1t} = P_t = 0 \tag{1.8}$$

$$m_2 V_{2n} - m_2 v_{2n} = -P_n \tag{1.9}$$

$$m_2 V_{2t} - m_2 v_{2t} = -P_t = 0 \tag{1.10}$$

Adding Eqs. 1.7 and 1.9 gives a single equation,

$$m_1 V_{1n} + m_2 V_{2n} = m_1 v_{1n} + m_2 v_{2n} \tag{1.11}$$

At this point, Eqs. 1.8, 1.10, and 1.11 provide three equations. The fourth equation is found by recognizing that in real impacts the deformation due to impact causes energy to be dissipated.

Two approaches can be used to obtain the last equation. For point-mass collisions, both approaches yield the same result. The first is conceptually to divide the duration of contact $\tau_2 - \tau_1$, into two parts. These are $\tau_1$ to $\bar{\tau}$ and $\bar{\tau}$ to $\tau_2$. During early contact from $\tau_1$ to $\bar{\tau}$, the masses approach and compress each other as their mass centers move closer together. Later during rebound, $\bar{\tau}$ to $\tau_2$, the mass centers move apart. The time $\bar{\tau}$ is when the relative normal velocity is zero. The normal impulse $P_n$ is also divided into two corresponding parts, $P_A$ during approach and $P_R$ during rebound, such that

$$e = P_R / P_A \tag{1.12}$$

where, $0 \leq e \leq 1$, and $e$ is called the (kinetic) coefficient of restitution. This concept is attributed to Newton.[7,8]

A second approach consists of a kinematic definition of the coefficient of restitution such that

$$V_{2n} - V_{1n} = -e(v_{2n} - v_{1n}) \tag{1.13}$$

where the velocity differences are the relative normal velocities of rebound and approach, respectively. It is now shown that the quantity $e$ in Eq. 1.13 is the same as the $e$ in Eq. 1.12. For the time interval $\tau_1$ to $\bar{\tau}$, Eqs. 1.7 and 1.9 give

$$m_1 \bar{v} - m_1 v_{1n} = P_A$$

and

$$m_2 \bar{v} - m_2 v_{2n} = -P_A$$

where it is noted that the masses $m_1$ and $m_2$ have the same normal velocity at $\tau = \bar{\tau}$. Similarly, for the time interval $\bar{\tau}$ to $\tau_2$

$$m_1 V_{1n} - m_1 \bar{v} = P_R$$

and

$$m_2 V_{2n} - m_2 \overline{v} = -P_R$$

Solving for $\overline{v}$ from each pair of equations gives, respectively,

$$\overline{v} = \frac{m_1 v_{1n} + m_2 v_{2n}}{m_1 + m_2} = \frac{m_1 V_{1n} + m_2 V_{2n}}{m_1 + m_2}$$

Using Eq. 1.12 and the preceding equations gives

$$m_1 V_{1n} - m_1 \overline{v} = e(m_1 \overline{v} - m_1 v_{1n})$$

Substituting for $\overline{v}$ and simplifying gives Eq. 1.13, where the velocity differences are the relative normal velocities of rebound and approach, respectively.

Equations 1.8, 1.10, 1.11, and 1.13 provide a set of four equations and four unknowns and can be solved. The solution provides the final velocities of a frictionless impact of two particles in a plane, which are

$$V_{1t} = v_{1t} \tag{1.14}$$

$$V_{2t} = v_{2t} \tag{1.15}$$

$$V_{1n} = v_{1n} + \frac{m_2}{m_1 + m_2}(1 + e)(v_{2n} - v_{1n}) \tag{1.16}$$

$$V_{2n} = v_{2n} - \frac{m_1}{m_1 + m_2}(1 + e)(v_{2n} - v_{1n}) \tag{1.17}$$

### ■ *Example 1.2*

PROBLEM: After ingestion into a turbine engine, a particle of sand bounces with velocities $v_{1n} = -30$ and $v_{1t} = 120$ m/s from the surface of a turbine blade. Another incoming particle with velocity components, $v_{2n} = 240$ m/s and $v_{2t} = 90$ m/s collides with the first particle. If $m_1 = 2.4 \times 10^{-12}$ kg, $m_2 = 3.1 \times 10^{-12}$ kg, and $e = 0.6$, find the final velocities of both particles. Calculate the total kinetic energy loss due to the collision as a fraction of the initial energy. Neglect friction.

*Solution:* Because of the frictionless nature of the impact, Eqs. 1.8 and 1.10 show that the final tangential velocities must equal the initial. That is,

$$V_{1_t} = v_{1_t} = 120 \text{ m/s}$$

and

$$V_{2_t} = v_{2_t} = 90 \text{ m/s}$$

Equations 1.11 and 1.13 become

$$2.4V_{1_n} + 3.1V_{2_n} = 2240 \text{ m/s}$$

and

$$V_{2_n} - V_{1_n} = -540 \text{ m/s}$$

The solution is $V_{1_n} = 213.5$ m/s and $V_{2_n} = 51.5$ m/s. The initial and final system kinetic energies are, respectively,

$$T_1 = \tfrac{1}{2}m_1(v_{1_n}^2 + v_{1_t}^2) + \tfrac{1}{2}m_2(v_{2_n}^2 + v_{2_t}^2) = 0.120 \times 10^{-6} \text{ N m}$$

and

$$T_2 = \tfrac{1}{2}m_1(V_{1_n}^2 + V_{1_t}^2) + \tfrac{1}{2}m_2(V_{2_n}^2 + V_{2_t}^2) = 0.886 \times 10^{-7} \text{ N m}$$

The fraction of energy lost is 0.262. ■

Since initial velocities typically are known and the final velocities are given by Eqs. 1.14 through 1.17, the loss of kinetic energy for a frictionless, two-particle impact can be determined. This is

$$T_L = \tfrac{1}{2}m_1 v_1^2 + \tfrac{1}{2}m_2 v_2^2 - \tfrac{1}{2}m_1 V_1^2 - \tfrac{1}{2}m_2 V_2^2$$

$$T_L = \tfrac{1}{2}\overline{m}(1 - e^2)(v_{2n} - v_{1n})^2 \tag{1.18}$$

where $\overline{m} = m_1 m_2/(m_1 + m_2)$. Note that for $e = 0$, the largest possible amount of energy is lost because of the impact, but not necessarily all of the initial energy. For $e = 1$, no energy is lost.

## ■ *Example 1.3*

PROBLEM: An experimental frontal collision of an automobile, $m_1$, into a rigid barrier is intended to reproduce approximately a direct frontal collision into another vehicle, $m_2$. The initial vehicle-to-vehicle crash velocities are $v_1$ and $-v_1$, respectively. Determine the equivalent barrier collision velocity, $v_b$, of $m_1$ to

a. dissipate the same energy as the vehicle-to-vehicle collision;
b. have the same momentum change as the vehicle-to-vehicle collision.

Assume perfectly plastic collisions and that the equivalent-energy collision dissipates one-half of the total energy of the vehicle-to-vehicle collision.

*Solution:* For the vehicle-to-vehicle collision, Eq. 1.18 gives

$$T_{Lv} = \tfrac{1}{2}\overline{m}(2v_1)^2$$

For the barrier collision

$$T_{Lb} = \tfrac{1}{2}m_1 v_b^2$$

Since $T_{Lb} = T_{Lv}/2$, the equivalent-energy barrier speed is

$$v_b = v_1\sqrt{\frac{2m_2}{m_1 + m_2}} \qquad \text{(a)}$$

For the vehicle-to-vehicle collision, the final velocity for vehicle 1 is, from Eq. 1.16,

$$V_{1n} = \frac{m_1 - m_2}{m_1 + m_2}$$

The corresponding linear momentum change is

$$m_1(V_{1n} - v_1) = \frac{-2m_1 m_2}{m_1 + m_2} v_1$$

This must equal the barrier impact momentum change, $-m_1 v_b$. Equating gives

$$v_b = v_1 \frac{2m_2}{m_1 + m_2} \quad \text{(b)}$$

Equations a and b give the equivalent barrier velocities for vehicle 1. ■

### 1.2.2 Collision of a Particle with a Flat Massive Plane

An important special case of the previous section occurs when the mass of one of the particles is significantly larger than that of the other, say $m_2 >> m_1$. Under this condition, the velocity change of $m_2$ is negligible and the solution is simpler. With only one particle, momentum is not conserved during the collision. Only two equations remain, Eqs. 1.8 and 1.13. These give the solution directly as

$$V_{1n} = -ev_{1n} \quad (1.19)$$

and

$$V_{1t} = v_{1t} \quad (1.20)$$

The particle rebounds from the surface with no change in tangential speed (frictionless case) and with a redirected and reduced normal velocity if $e < 1$. For the perfectly elastic case, $e = 1$, the rebound is a mirror image of the approach.

■ *Example 1.4*

PROBLEM: A hockey puck as shown in the illustration strikes the ice with its flat surface at point $A$. It bounces and follows the trajectory $ABC$. If the maximum height, at $B$, is 1.3 in., and $AC$ is 26 in., determine the puck's velocity prior to impact at $A$. Use $e = 0.2$.

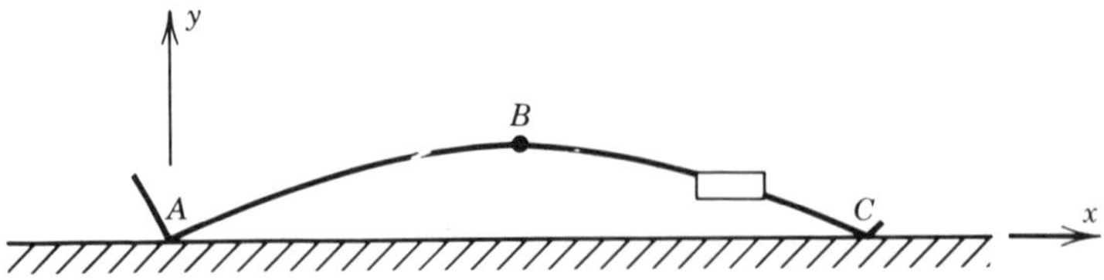


**Figure E1.4.** Bouncing puck.

*Solution:* If air resistance is neglected, the trajectory from $A$ to $C$ is parabolic with the equation

$$y = \frac{v_{y0}}{v_{x0}} x - \frac{g}{2v_{x0}^2} x^2$$

where $g = 386$ in./s$^2$, $v_{x0}$ and $v_{y0}$ are the initial trajectory velocities (and also the final impact velocities). Two sets of trajectory coordinates are known. These are $(x_B, y_B) = (13, 1.3)$ and $(x_c, y_c) = (26, 0)$. These provide the solution for the two velocities using the trajectory equation. These are $v_{x0} = 158.5$ in./s and $v_{y0} = 31.7$ in./s. From Eqs. 1.19 and 1.20, the velocity components at $A$ prior to impact are

$$V_{x0} = v_{x0} = 158.5 \text{ in./s}$$

and

$$v_{y0} = -\frac{V_{y0}}{e} = -158.5 \text{ in./s}$$ ■

## 1.3 COMMENTS AND DISCUSSION

The last two sections briefly present the elementary impact equations typically introduced in dynamics courses to model collisions of particles. These equations are very simple yet provide insight into the effect the impact process has on rigid body motion. A simplistic view is taken for the local dynamic behavior of the colliding objects. During the initial part of contact, perpendicular or normal compressive deformation occurs and is at least partially returned during rebound; see Fig. 1.3. Real collisions lose energy and are accompanied by a reduced rebound velocity

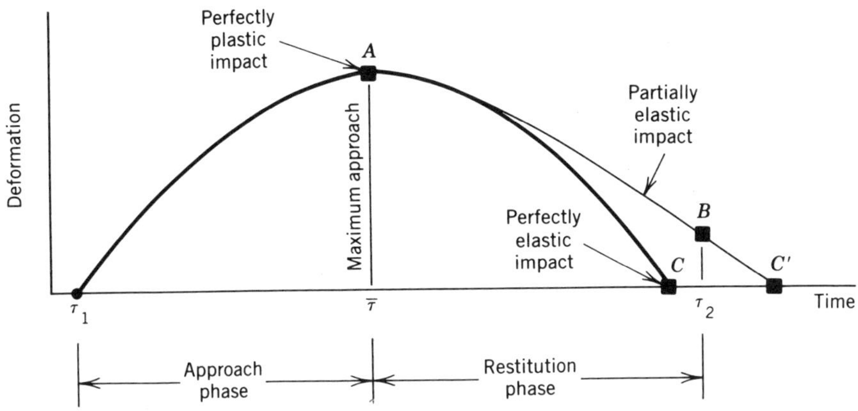


**Figure 1.3.** Deformation history of an impact.

and possibly permanent deformation (points $A$ or $B$, for example) or by no permanent deformation (point $C'$). An ideal impact with no energy loss ends at point $C$. The corresponding interparticle normal force varies with time, but only its impulse (its integral) needs to be considered, since only velocity changes are calculated. The intent of the classical approach to impact is to permit calculation of final velocities with the presumption that the initial velocities are known.

To calculate the final velocities, more information than the initial velocities is needed. The equations contain the mass values and the coefficient of restitution. This coefficient is treated as a constant in the impact equations because it does not depend on the unknowns (the final velocities). However, in practice, the restitution coefficient does depend on the particle shapes, materials, and initial velocities. Values must be determined experimentally or estimated analytically.[8-10] For most engineering applications $e$ varies with relative initial normal velocities as shown in Fig. 1.4.

An "inverse problem" often arises in practice. This situation occurs when the final velocities are known and the corresponding initial velocities are desired. In fact, many combinations of knowns and unknowns can occur. A mixed problem is possible where some initial and some final velocity components are known and the others are unknown. In experimental work, all the velocity components may be known and the coefficient of restitution or the masses are to be determined. In general, Eqs. 1.8, 1.10, 1.11, and 1.13 may be considered as four equations that

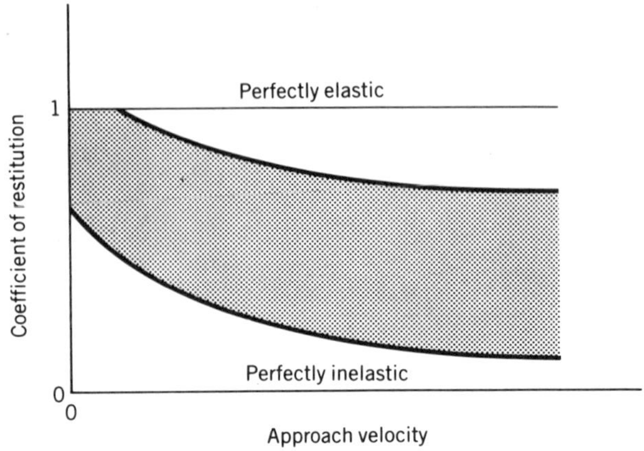


**Figure 1.4.** Common trend of coefficients of restitution.

can be used to calculate a variety of combinations of any four unknowns. Depending on the particular combination of knowns and unknowns, these equations may be linear or nonlinear. Furthermore, in practice, the coefficient of restitution depends on the initial velocities. It may be necessary to take this dependency into account in inverse solutions and when dealing with experimental data.

In addition to knowledge of the initial velocities, the masses, and the coefficient of restitution, the assumptions made in Section 1.2.1 must be satisfied. In order to discuss their meaning, consider an example of the collision of a baseball with a bat. (Many of the assumptions may need to be relaxed to be able to analyze this problem accurately, such as the planar restriction, small angular velocities, small deformation, and that of smooth surfaces. These assumptions will be removed later in this book. Sliding of an initially spinning ball on the bat may or may not continue through separation. The conditions of large forces and short contact durations must always remain. These are not usually restrictive, however, since they form the definition of an impact.) The duration of contact between the baseball and bat is small compared with the ball's times of flight, both from the pitcher's hand to the bat and from the bat to the bleachers. The contact force between the ball and bat is orders of magnitude greater than the ball's weight (gravity force); neglecting gravity over the contact duration is usually an accurate assumption.[15] The same would be true for the force between an automobile's tires and ground during a high-speed vehicle collision. Another example is the bouncing of a rubber ball against a wall (gravity force is tangential) and bouncing against a floor (gravity force is normal). The same impact analysis may apply to both cases.

It should be pointed out that the primary goal of this book is to develop and illustrate methods that allow the calculation of changes in rigid body motion, specifically velocity changes, due to an impact. These methods are intended to provide simple models of the impact process applicable to dynamical simulations. These simulations may be of the impact process itself (such as a bouncing ball for computer animation) or that of a complex system analysis where the impact plays a reduced role (such as a feedback-controlled robotic system). It is not intended to study the local deformations and material behavior that occur during an impact. The study of the local, dynamic elastic/plastic behavior during an impact can be an extremely difficult and complicated problem, and some excellent books are available that cover this topic.[8, 12–14] However, for dynamic simulations the local behavior often can be represented using coefficients such as the coefficient of restitution and others that will be introduced later. These coefficients can be determined analytically or experimentally or both. When known, either accurately or approximately, the results of this book permit the velocity changes to be calculated. These velocity components are those of the mass center. Vibrations result from collisions; dynamic elastic motion distributed throughout an object during and following a collision can be quite extensive. None of this is included in this approach. In some cases, vibrations may decay quickly, even before separation occurs. Other times they may not. In all cases, the mass center velocity changes can be calculated. Without studying local behavior, forces and stresses are also not determined. In some problems, distributed elastic dynamics and rigid body dynamics must be combined to provide a comprehensive understanding of the impact process.

## REFERENCES


1 Beer, F. P. and Johnson, R. J., Jr., *Mechanics for Engineers, Statics and Dynamics*, McGraw-Hill, New York, 1976.

2 Goodman, L. E. and Warner, W. H., *Statics and Dynamics*, Wadsworth, Belmont, CA 1964.

3 Higdon, A., et al., *Engineering Mechanics*, Prentice-Hall, Englewood Cliffs, NJ, 1979.

4 Meriam, J. L., *Engineering Mechanics, Statics and Dynamics*, John Wiley, New York, 1978.

5 Sandor, B. I., *Engineering Mechanics, Statics and Dynamics*, Prentice-Hall, Englewood Cliffs, NJ, 1983.

6 Pestel, E. C. and Thomson, W. T., *Dynamics*, McGraw-Hill, New York, 1968.

7 Timoshenko, S. and Young, D. H., *Advanced Dynamics*, McGraw-Hill, New York, 1948.

8 Goldsmith, W., *Impact*, Edw. Arnold, London 1960.

9 Koller, M. G. and Busenhart, M., "Elastic Impact of Spheres on Thin Shallow Spherical Shells," *Int. J. Impact Eng.*, **4**(1), 11–22, 1986.

10 Barkan, P., "Impact," *Mechanical Design and Systems Handbook*, Section 16, McGraw-Hill, New York, 1974.

11 Brach, R. M., "Impact Dynamics with Applications to Solid Particle Erosion," *Int. J. of Impact Eng.*, **7**(1), 37–54, 1988.

12 Zukas, J. A., *Impact Dynamics*, John Wiley, New York, 1982.

13 Johnson, W., *Impact Strength of Materials*, Edw. Arnold, London 1972.

14 Kolsky, H., *Stress Waves in Solids*, Dover, New York, 1963.

15 Adair R. K., *The Physics of Baseball*, Harper & Row, New York, 1990.

# CHAPTER 2

# POINT-MASS COLLISIONS IN A PLANE

The equations developed in this chapter and their further interpretation and examination in the next chapter form a basis of mechanical impact theory as covered in this book. Later chapters treat more complicated conditions such as rigid body impacts and specific applications. Tangential effects, which were neglected in Chapter 1, are now included. The overall problem remains the same, that is, to find the final velocities following a planar collision of two point masses where the initial velocities are known. Again, the coefficient of restitution is used. Another constant, $\mu$, is defined, where $\mu$ **is the ratio of the tangential and normal impluse components.** This definition of $\mu$ represents what appears to be a subtle difference from other treatments of the impact process,[1] where $\mu$ typically is treated as a coefficient of friction. It will become apparent later that this distinction can be crucial in the proper modeling of collisions.

The final velocity components are found from the solution of four linear algebraic equations. The constants $e$ and $\mu$ form two parameters in the solution equations. Their values must be found by calculation or experiment or both for specific applications.

In this and later chapters, the equations that are obtained from the principles of impulse and momentum and those obtained from physical processes (such as restitution) will be called the *system equations.* Their solution set, which typically provides the final velocities, will be referred to as the *solution equations*.

## 2.1 SYSTEM EQUATIONS (EQUATIONS OF IMPULSE AND MOMENTUM)

Figure 1.2 shows free body diagrams of two colliding particles along with an orthogonal reference system with coordinates $n$, $t$, and $t'$. A two-dimensional, central oblique collision will be analyzed here such that velocity components in the $t'$ direction do not exist either before or after the collision. The coordinate system is chosen such that the $t$–$t'$ plane is perpendicular to the line of mass centers and all initial velocities lie in the $n$–$t$ plane. The quantities $P_n$ and $P_t$ in Fig. 1.2 represent orthogonal components of the impulse of the interparticle contact force. Their resultant has the magnitude

$$P = \sqrt{P_n^2 + P_t^2} \tag{2.1}$$

Applying the principle of impulse and momentum to each particle and in each of the two coordinate directions gives four equations.* For the first particle,

$$m_1 V_{1n} - m_1 v_{1n} = P_n \tag{2.2.1}$$

$$m_1 V_{1t} - m_1 v_{1t} = P_t \tag{2.2.2}$$

For the second particle,

$$m_2 V_{2n} - m_2 v_{2n} = -P_n \tag{2.3.1}$$

$$m_2 V_{2t} - m_2 v_{2t} = -P_t \tag{2.3.2}$$

Addition of pairs of Eqs. 2.2 and 2.3 eliminates the impulses $P_n$ and $P_t$ and gives an expression of conservation of momentum in each coordinate direction:

$$m_1 V_{1n} + m_2 V_{2n} = m_1 v_{1n} + m_2 v_{2n} \tag{2.4}$$

$$m_1 V_{1t} + m_2 V_{2t} = m_1 v_{1t} + m_2 v_{2t} \tag{2.5}$$

*Throughout this book, final velocities are capitalized; initial, or precollision, velocities are written in lower case.

Equations 2.4 and 2.5 do not contain the impulse coordinates $P_n$ and $P_t$. Additional equations containing only velocity variables are needed. These are ultimately solved; $P_n$ and $P_t$ can then be calculated using Eq. 2.2.2–2.3.2.

According to Newton's law as expressed in Chapter 1, $P_n$ and $P_t$ must be components of the resultant impulse of all forces external to a particle. Here, the total impulse $P$ is attributed to the interparticle force, that is, the force due only to the collision itself. This represents an assumption typically made when studying collisions. The impact force is assumed to be large; impulses due to all other forces are assumed to be small and are neglected. The validity of this assumption depends on the specific application. The effect of external impulses can be included, however, and is taken up later. Note that no assumptions have been made yet concerning the duration of particle contact. Strictly speaking, Eqs. 2.2–2.5 are valid for any time duration. When applied to the study of collisions, an assumption of short contact durations is made in order to discount certain other effects, however. For rigid body impacts (discussed in a later chapter), short durations are assumed in order to be able to neglect changes in orientation and any associated geometry changes of the two objects during contact. In effect, the short-duration assumption corresponds to infinitesimal displacements and finite velocity changes. This, of course, implies large accelerations, which corresponds to the earlier assumption of a large interparticle impact force.

Equations 2.4 and 2.5 provide a pair of linear equations containing eight velocity components of the system of two particles. In a typical problem formulation the initial velocities are known, and the four final velocity components are unknown and are to be calculated. Two additional equations are necessary for a solution. A typical approach is to relate the relative normal velocity components by means of the coefficient of restitution, $e$, such that

$$V_{2n} - V_{1n} = -e(v_{2n} - v_{1n}) \tag{2.6}$$

where $0 \leq e \leq 1$. Equation 2.6 requires that the relative normal velocity of the two masses following impact, $V_{2n} - V_{1n}$, must be zero, equal to, or some fraction of the relative normal velocity of approach, $v_{2n} - v_{1n}$. This corresponds to observed behavior. Equation 2.6 is a kinematic constraint with $e$ as the controlling parameter. For example, setting $e$ to zero forces $V_{2n}$ to equal $V_{1n}$. Equally important, however, the coefficient $e$

accounts for energy lost during the collision. That is, Eq. 2.6 is an alternative to the expression

$$\tfrac{1}{2}m_1(V_{1t}^2 + V_{1n}^2) + \tfrac{1}{2}m_2(V_{2t}^2 + V_{2n}^2) + T_L$$
$$= \tfrac{1}{2}m_1(v_{1t}^2 + v_{1n}^2) + \tfrac{1}{2}m_2(v_{2t}^2 + v_{2n}^2) \qquad (2.7)$$

where $T_L$ is the kinetic energy loss. When solving for final velocities, Eq. 2.6 is preferred to 2.7 because 2.6 is linear with respect to the unknown velocity components and permits a simpler solution. There is no advantage in using Eq. 2.7 as a system equation, since a unique relationship exists between $T_L$ and the coefficients as shown in the next chapter.

One other equation is needed to provide the four required for the solution of the collision problem. Tangential effects must be considered; these will influence the changes of velocity components in the tangential, $t$, direction. For an oblique impact when the particles have a nonzero initial relative tangential velocity component, sliding must always occur, at least during initial stages of contact. A question that must be answered is whether or not relative tangential motion continues through separation or ceases prior to separation. Although the laws of Coulomb friction do not apply during all impacts,[2] the concept and mathematical form of a coefficient of friction prove to be very convenient. Tangential effects can be treated by use of a constant $\mu$ such that

$$\mu = P_t/P_n \qquad (2.8)$$

Although $\mu$ is an arbitrary constant, the tangential force and its impulse usually are found to be dissipative, such as from friction or plastic deformation, and cannot add energy to the system. For a particle, the tangential impulse always opposes initial relative tangential motion. Consequently, for point-mass collisions, $\mu$ can be either positive or negative with its sign given by

$$\operatorname{sgn}(\mu) = \operatorname{sgn}\left(\frac{v_{2t} - v_{1t}}{v_{2n} - v_{1n}}\right) \qquad (2.9)$$

When relative tangential motion does not exist at separation, then

$$V_{1t} = V_{2t} \tag{2.10}$$

is true. When modeling tangential effects, Eqs. 2.8 and 2.10 are often viewed as alternatives, appropriately chosen to model the collision. Selection of these alternatives will be discussed in much greater depth in the next chapter. Equation 2.8 states simply that $\mu$ is a constant equal to the ratio of impulse components. In specific applications the appropriate value of $\mu$ must be determined experimentally or estimated by means such as analytical modeling of the mechanism of tangential force generation. If Coulomb friction is appropriate, then $\mu$ is related to the coefficient of dynamic friction. Otherwise, it can be considered to be an equivalent coefficient of friction. Before Eq. 2.8 can be used to solve for velocities, the impulses must be eliminated. Use of Eqs. 2.2 and 2.3, allows Eq. 2.8 to be written as

$$m_1 V_{1t} - \mu m_2 V_{2n} = m_1 v_{1t} - \mu m_2 v_{2n} \tag{2.11}$$

## 2.2 SOLUTION OF SYSTEM EQUATIONS

Equations 2.4–2.6 and 2.11 or Eqs. 2.4–2.6 and 2.10 can be treated as four linear equations with four unknowns: $V_{1t}$, $V_{1n}$, $V_{2t}$, and $V_{2n}$. The normal impact equations, 2.4 and 2.6, can be solved independently and give

$$V_{1n} = v_{1n} + m_2(1 + e)(v_{2n} - v_{1n})/(m_1 + m_2) \tag{2.12}$$

$$V_{2n} = v_{2n} - m_1(1 + e)(v_{2n} - v_{1n})/(m_1 + m_2) \tag{2.13}$$

If sliding exists at separation, Eqs. 2.5 and 2.11 provide

$$V_{1t} = v_{1t} + \mu m_2(1 + e)(v_{2n} - v_{1n})/(m_1 + m_2) \tag{2.14}$$

$$V_{2t} = v_{2t} - \mu m_1(1 + e)(v_{2n} - v_{1n})(m_1 + m_2) \tag{2.15}$$

If sliding does not exist at separation, Eq. 2.10 is used instead of Eq. 2.8. The expressions for the tangential velocities can then be written as

$$V_{1t} = v_{1t} + m_2(v_{2t} - v_{1t})/(m_1 + m_2) \tag{2.16}$$

$$V_{2t} = v_{2t} - m_1(v_{2t} - v_{1t})/(m_1 + m_2) \tag{2.17}$$

The solution equations for $V_{1t}$ and $V_{2t}$ can be written in several ways; the forms chosen and used previously allow the change in tangential velocities to be seen by inspection and also relate the final velocities to the initial relative normal velocities in anticipation of a later use. The change in velocity components is obtained directly from Eqs. 2.12–2.15. In applications where one mass, say $m_2$, represents a moving massive barrier, the preceding equations still apply. It is necessary to let $m_1/m_2$ approach zero and consider $V_{2n} = v_{2n}$ and $V_{2t} = v_{2t}$.

Frequently, the change in magnitude of the velocity vector is needed. The vector velocity change for each particle is

$$\Delta v_i = \sqrt{\Delta v_{in}^2 + \Delta v_{it}^2}$$

$$\Delta v_i = \sqrt{(1 + e)(1 + \mu^2)}\ (\overline{m}/m_i)(v_{2n} - v_{1n}), \qquad i = 1, 2 \tag{2.18}$$

The final relative tangential velocity can be expressed in a concise manner in the following way. Let the relative velocities be given by

$$v_{tR} = v_{2t} - v_{1t} \tag{2.19.1}$$

$$v_{nR} = v_{2n} - v_{1n} \tag{2.19.2}$$

and

$$V_{tR} = V_{2t} - V_{1t} \tag{2.19.3}$$

Subtracting Eq. 2.14 from Eq. 2.15 then gives

$$V_{tR} = v_{tR} - \mu(1 + e)v_{nR} \tag{2.20}$$

This can be viewed as a linear equation relating $V_{tR}$ and $v_{tR}$ with a slope of unity and the second term as the intercept. This, illustrated in Fig.

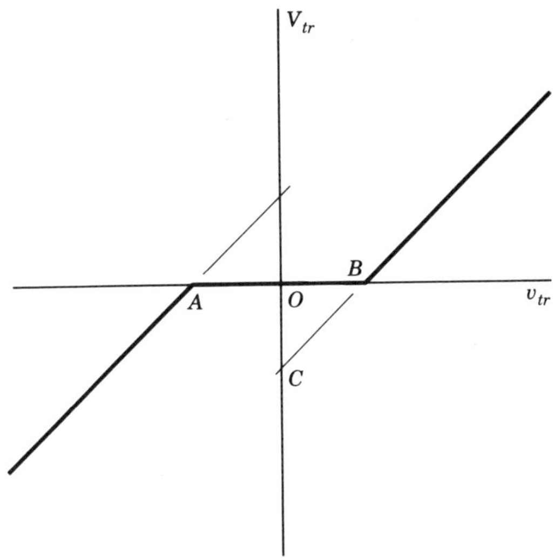


**Figure 2.1.** Relative tangential velocity; sliding ends prior to separation only for solutions between points *A* and *B*.

2.1, shows that the sliding solutions lie to the right of point *B* and to the left of point *A* (for negative $\mu$). Solutions corresponding to Eqs. 2.16 and 2.17 lie between points *A* and *B*. Point *C* is the intercept with the value $-\mu(1 + e)v_{nR}$. The value of the impulse ratio $\mu$ corresponding to points *A* and *B* can be found from Eq. 2.20 with $V_{tR} = 0$. The significance of this is considered in Chapter 3 for particles and Chapter 6 for rigid bodies.

## 2.3 SPECIAL PROPERTIES OF THE SYSTEM AND SOLUTION EQUATIONS

Although the four system equations solved previously fall into the general classification of simultaneous linear algebraic equations, they have a particular form that allowed them to be solved in sequential pairs. Equations 2.4 and 2.6 were solved to determine completely the normal velocity changes. Equation 2.2.1 or 2.3.1 then furnished the normal impulse $P_n$. Equations 2.5 and 2.11 (or 2.10) then were solved for the tangential velocity changes, which, in turn, determined $P_t$. In addition to the algebraic convenience of solving pairs of equations sequentially, this form of equations illustrates a characteristic of impact system equations that carries through into the rigid body and three-dimensional problems. From a physical perspective this mathematical feature shows that a tangential impulse cannot develop unless a normal impulse exists first. A normal collision (direct central impact) can occur with no tangential effects, but a tangential impact cannot occur without a normal component.

Because computer solutions are common, the equations treated as a simultaneous set are now reviewed.

### 2.3.1 Matrix Form of System Equations

Equations 2.4–2.6 and 2.11 form a set of four linear equations with four unknowns. These can be expressed in matrix form as

$$A\mathbf{V} = C\mathbf{v} \tag{2.21}$$

where $\mathbf{V}$ and $\mathbf{v}$ are vectors whose transposes can be written as $\mathbf{V}^T = \{V_{1n}\ V_{1t}\ \ V_{2n}\ \ V_{2t}\}$ and $\mathbf{v}^T = \{v_{1n}\ v_{1t}\ v_{2n}\ v_{2t}\}$; matrices $A$ and $C$ are

$$A = \begin{bmatrix} m_1 & 0 & m_2 & 0 \\ 0 & m_1 & 0 & m_2 \\ -1 & 0 & +1 & 0 \\ 0 & m_1 & -\mu m_2 & 0 \end{bmatrix}, \qquad C = \begin{bmatrix} m_1 & 0 & m_2 & 0 \\ 0 & m_1 & 0 & m_2 \\ e & 0 & -e & 0 \\ 0 & m_1 & -\mu m_2 & 0 \end{bmatrix}$$

A change of unknowns from final velocities to velocity-change variables produces a set of equations with simplified solution properties. Let $\mathbf{U} = \mathbf{V} - \mathbf{v}$, then, using Eq. 2.21,

$$A\mathbf{U} = A\mathbf{V} - A\mathbf{v} = C\mathbf{v} - A\mathbf{v}$$

or

$$A\mathbf{U} = B\mathbf{v} \tag{2.22}$$

where $B$ is the difference of the $C$ and $A$ matrices,

$$B = \begin{bmatrix} 0 & 0 & 0 & 0 \\ 0 & 0 & 0 & 0 \\ (1+e) & 0 & -(1+e) & 0 \\ 0 & 0 & 0 & 0 \end{bmatrix}$$

Matrix forms can be useful when impact capability is to be embedded into a set of model equations representing the many elements of a large system.

### 2.3.2 Symmetry of Equations: The Inverse Problem

Several properties of the preceding system equations occasionally can be exploited. A certain symmetry exists between each final velocity component, such as $V_{1n}$, and its initial counterpart, $v_{1n}$. (See matrices $A$ and $C$). An interchange of initial with final velocity symbols affects only Eq. 2.6. But replacing $e$ by $1/e$ "corrects" this equation. The implication is that if final velocities are known and the initial velocity components are unknown (the inverse problem), the latter can be obtained from the solution Eqs. 2.12–2.15, simply by replacing $e$ by its reciprocal and interchanging initial and final velocity symbols. This cannot be done when $e = 0$, which is a singular point in the inverse solution.

### 2.3.3 Coefficient of Restitution: Negative Values

It is typically stated that the coefficient of restitution is bounded by 0 and 1 owing to the nature of its definition and to physical reasoning. Consider, however, the case of an errant baseball passing through a window screen. Mechanical failure of the screen wire removes energy from the ball and its velocity is lower following penetration. The process of penetration such as the baseball screen example can be modeled by the same equations already presented by permitting $e$ to take on values between 0 and $-1$. In the next chapter it is shown that energy loss depends on $e^2$, and negative values of $e$ pose no fundamental difficulty. Equation 2.6 is easily interpreted for negative $e$, since velocity reversal does not occur.

In addition to $-1 < e \leq 0$, the value of $e = -1$ has an interesting implication, which will prove useful later when developing a solution to the rigid body collision problem. Substitution of $V_{2n} - v_{2n}$ from Eqs. 2.13 into 2.3 and using Eq. 2.6 gives

$$eP_n = -(1 + e)\overline{m}(V_{2n} - V_{1n}) \tag{2.23.1}$$

$$P_n = (1 + e)\overline{m}(v_{2n} - v_{1n}) \tag{2.23.2}$$

where $\overline{m} = m_1 m_2/(m_1 + m_2)$. Using physical reasoning it can be argued that a tangential impulse $P_t$ cannot develop unless $P_n$ exists. Furthermore, using Eqs. 2.8 and 2.1, the magnitude of the total impulse, $P$, is

$$P = \sqrt{1 + \mu^2}\,P_n \tag{2.24}$$

Eq. 2.23.2 shows that $P_n$ will be zero when $v_{2n} = v_{1n}$, which in turn implies no impact. However, notice that the value of $e = -1$ (the seemingly trivial case of forceless penetration) also has the effect of forcing $P_n$ and, consequently, $P$ to be 0. The significance of this observation is exploited in Chapter 5 when analyzing impacts of rigid bodies.

### 2.3.4 Pulse Shapes: Peak Force

By the nature of the mechanical impact process, the interparticle force is nonzero only during the time interval of contact, $\tau_1$ to $\tau_2$. Figure 2.2 shows a typical force–time pulse. It is frequently convenient to use a triangular force with the actual impulse as an idealization,[3] as shown dashed in Fig. 2.2. By definition, the normal impulse is the area under the force–time curve, or

$$P_n = F_p(\tau_2 - \tau_1)/2 \tag{2.25}$$

Using Eq. 2.23.2 provides a relationship between peak and duration of the idealized force and the initial normal velocities and the coefficient at restitution. That is,

$$F_p = 2\overline{m}(1 + e)(v_{2n} - v_{1n})/(\tau_2 - \tau_1) \tag{2.26}$$

This shows that compared to a perfectly inelastic impact ($e = 0$), a perfectly elastic impact ($e = 1$) will have twice the impulse and, for the

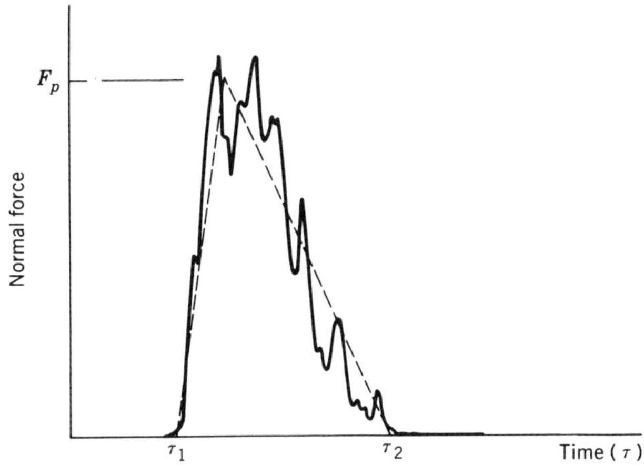


**Figure 2.2.** Typical normal force variation for an impact.

same pulse duration, twice the peak force. Examples of typical pulse shapes are available from many sources. For example, Carden[3] compares different shapes for aircraft–ground impacts and concludes that triangular shapes are good approximations for that application.

## ■ *Example 2.1*

PROBLEM: Two vehicles collide at right angles as shown. Vehicle 2 is initially at rest; vehicle 1 moves straight ahead at 30 mph into vehicle 2. Following the impact, both move with the same speed in the original direction of vehicle 1. The graph shows an experimental acceleration record of the mass center of vehicle 1 from this type of collision. Use this curve and the dashed triangular pulse to estimate the peak deceleration of vehicle 1 and the peak intervehicular force during the impact. Vehicle weights are $w_1 = 4900$ lb; $w_2 = 2256$ lb.

*Solution:* Since both vehicles have the same final speed, $e = 0$. Equation 2.12 gives

$$V_{1n} = V_{2n} = v_{1n} + \frac{m_2}{m_1 + m_2}(1 + e)(v_{2n} - v_{1n})$$

$$= -44 + 13.9 = -30.1 \text{ ft/s}$$

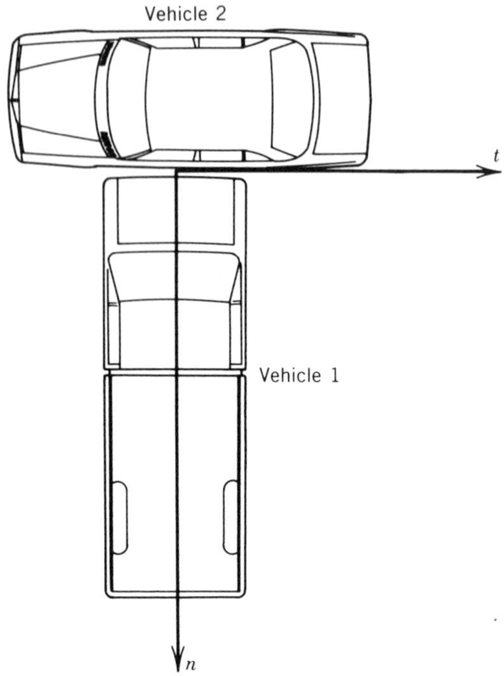


**Figure E2.1.1.** Vehicle orientation at initiation of impact.

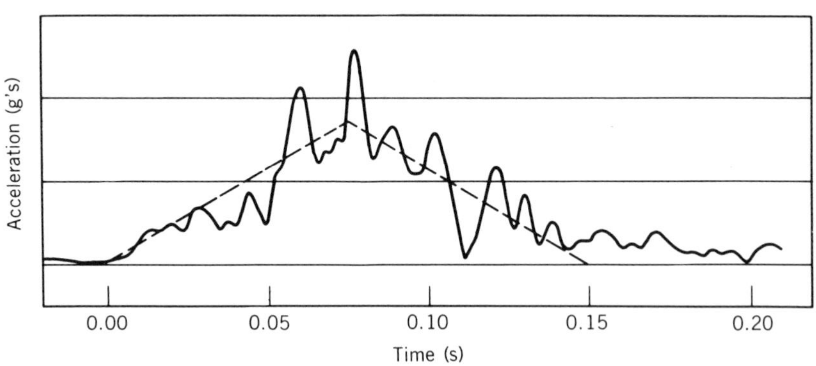


**Figure E2.1.2.** Lateral acceleration of vehicle 2.

The idealized, triangular peak force is obtained using Eq. 2.26. Using $\tau_2 - \tau_1 = 0.15$ s, the curve gives

$$
\begin{aligned}
F_p &= 2\overline{m}(1 + e)(v_{2n} - v_{1n})/(\tau_2 - \tau_1) \\
&= 2(48)(1 + 0)(44)/(0.15) \\
F_p &= 28{,}160 \text{ lb}
\end{aligned}
$$

The corresponding peak deceleration (triangular pulse) of vehicle 1 is 5.75 g. The actual peak of the acceleration curve is about 50% higher than the triangular peak. This gives an actual maximum force of about 42,200 lb.

The velocity change of vehicle 1 is 13.9 ft/s; this gives an impulse of 2115 lb-s. For a time interval of 0.15 s, this gives an average force of 14,100 lb. A rule of thumb for vehicle impacts is that the maximum force is two to three times the average force. This is 28,200 to 42,300 lb, and includes the value of 42,200 lb computed previously. ■

### 2.3.5 Motion of the Mass Center of the System

Since the effect of all forces external to the system of two particles is neglected, the velocity of the mass center must remain constant during the collision (or as long as external forces can be neglected). These velocity components are

$$\overline{v}_n = \overline{m}\left(\frac{v_{1n}}{m_2} + \frac{v_{2n}}{m_1}\right) \tag{2.27.1}$$

and

$$\overline{v}_t = \overline{m}\left(\frac{v_{1t}}{m_2} + \frac{v_{2t}}{m_1}\right) \tag{2.27.2}$$

## 2.4 TANGENTIAL RESTITUTION

There is some valid argument and also experimental evidence[4] to indicate that significant shearing effects, or what might be termed tangential elastic effects, exist during some impacts. While this is discussed more fully in Chapter 6, the equations for tangential restitution are presented here. This is when the initial relative tangential velocity is reversed and can be modeled with the equation

$$V_{2t} - V_{1t} = -e_t(v_{2t} - v_{1t}) \tag{2.28}$$

This equation has a form identical to Eq. 2.6, which models normal restitution. The quantity $e_t$ is a coefficient of tangential restitution. Equations 2.4–2.6 and 2.28 now become the system equations. Their solution produces Eqs. 2.12 and 2.13 for the final normal velocity components. The final tangential components are

$$V_{1t} = v_{1t} + m_2(1 + e_t)(v_{2t} - v_{1t})/(m_1 + m_2) \tag{2.29}$$

$$V_{2t} = v_{2t} - m_1(1 + e_t)(v_{2t} - v_{1t})/(m_1 + m_2) \tag{2.30}$$

Equations 2.8 and 2.28 represent different physical effects, tangential retardation and a form of tangential elasticity, respectively. But recall that Eq. 2.8 states simply that $\mu$ is the ratio of the tangential to normal impulse components, $P_t/P_n$, *regardless of the mechanism that generates the tangential effects.* Comparison of the solution Eqs. 2.14 and 2.15 with the solution Eqs. 2.29 and 2.30 shows that if a tangential restitution model is used,

$$\mu = \frac{1 + e_t}{1 + e} \frac{v_{2t} - v_{1t}}{v_{2n} - v_{1n}} \tag{2.31}$$

and both solutions can provide the same results. In summary, when tangential restitution (tangential velocity reversal) occurs, the corresponding coefficient, $e_t$, and Eq. 2.28 can be used as an appropriate system equation. However, as was discussed previously for the coefficient $e$, negative values of a restitituion coefficient can be interpreted as retardation rather than rebound. Examination of Eq. 2.31 shows that values of $e_t$ between $-1$ and 0 have a direct correspondence to values of $\mu$. In particular, $e_t = -1$ implies no tangential effects, that is, $\mu = 0$. Although solutions using $e_t$ and $\mu$ can be equivalent, it should be kept in mind that they represent different physical effects such as elasticity and friction. This topic is discussed in more detail in Chapter 6.

Some researchers[5] use Eq. 2.28 without the negative sign to define $e_t$. In this form, the coefficient serves as a parameter to illustrate experimentally measured tangential velocity changes. This gives positive values of $e_t$ when tangential restitution is not present and negative values when it is.

## 2.5 EFFECT OF EXTERNAL IMPULSES

Impulses of forces other than the interparticle contact forces are not always negligible. If they are known or can be estimated, they can be taken into account. Equations 2.2.1–2.3.2 can be rewritten as

$$m_1 V_{1n} - m_1 v_{1n} = P_n + P_{1n} \tag{2.32.1}$$

$$m_1 V_{1t} - m_1 v_{1t} = P_t + P_{1t} \tag{2.32.2}$$

$$m_2 V_{2n} - m_2 v_{2n} = -P_n + P_{2n} \tag{2.33.1}$$

and

$$m_2 V_{2t} - m_2 v_{2t} = -P_t + P_{2t} \tag{2.33.2}$$

where $P_{in}$ and $P_{it}$ are impulse components of external forces on mass $i$, $i = 1, 2$. Adding pairs of these equations to eliminate $P_n$ and $P_t$ gives

$$m_1 V_{1n} + m_2 V_{2n} = m_1 v_{1n} + m_2 v_{2n} + P_{1n} + P_{2n} \tag{2.34}$$

and

$$m_1 V_{1t} + m_2 V_{2t} = m_1 v_{1t} + m_2 V_{2t} + P_{1t} + P_{2t} \tag{2.35}$$

These replace the earlier conservation of momentum equations and account for the fact that momentum is no longer conserved. Equation 2.6, which contains the restitution coefficient, $e$, and Eq. 2.8, which defines the impulse ratio $\mu$, are both still valid. Therefore, Eqs. 2.6, 2.8, 2.34, and 2.35 form a set of system equations. Their solution is found to be

$$V_{1n} = v_{1n} + \frac{\overline{m}}{m_1}(1 + e)(v_{2n} - v_{1n}) + \frac{P_{1n} + P_{2n}}{m_1 + m_2} \tag{2.36}$$

$$V_{2n} = v_{2n} - \frac{\overline{m}}{m_2}(1 + e)(v_{2n} - v_{1n}) + \frac{P_{1n} + P_{2n}}{m_1 + m_2} \tag{2.37}$$

$$V_{1t} = v_{1t} + \mu \frac{\overline{m}}{m_1}\left[(1 + e)(v_{2n} - v_{1n}) - \frac{P_{1n}}{m_1} + \frac{P_{2n}}{m_2}\right] + \frac{P_{1t}}{m_1} \quad (2.38)$$

$$V_{2t} = v_{2t} - \mu \frac{\overline{m}}{m_2}\left[(1 + e)(v_{2n} - v_{1n}) - \frac{P_{1n}}{m_1} + \frac{P_{2n}}{m_2}\right] + \frac{P_{2t}}{m_2} \quad (2.39)$$

For low-speed collisions, the interparticle force may be small and the impulse of other forces such as weight may not be negligible. To calculate the impulse of other forces, the duration of the collision must be known; see Eq. 1.3.

## ■ *Example 2.2*

PROBLEM: During the vehicle impact of Example 2.1, the wheels of vehicle 2 slide sideways with a road–tire friction coefficient $f = 0.75$. Assuming that the normal forces between vehicle 2 and the pavement remain constant, compare the velocity change of vehicle 1 with and without the frictional effect of the sliding of the tires during the impact.

*Solution:* All external impulses are zero with the exception of $P_{2n}$. For constant friction and normal forces,

$$P_{2n} = fw_2(\tau_2 - \tau_1) = 253.8 \text{ lb-s}$$

The velocity change of vehicle 1 is

$$\Delta V_{1n} = V_{1n} - v_{1n} = \frac{\overline{m}}{m_1}(1 + e)(v_{2n} - v_{1n}) + \frac{253.8}{m_1 + m_2}$$

$$\Delta V_{1n} = 13.9 + 1.1 = 15 \text{ ft/s}$$

Without the frictional effect, $\Delta V_{1n} = 13.9$ ft/s. Tire friction here causes about a 7.9% larger $\Delta V_{1n}$. ■

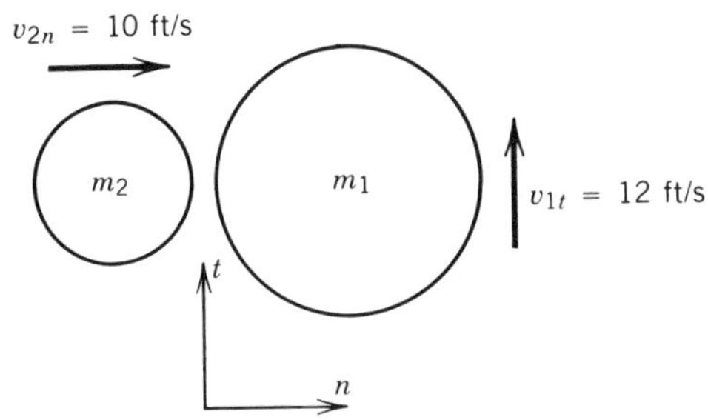


**Figure P2.2.** Point masses.

## 2.6 PROBLEMS

**2.1** Show that sliding (relative tangential motion) ends at separation when

$$\mu = \frac{1}{1 + e} \frac{v_{2t} - v_{1t}}{v_{2n} - v_{1n}}$$

**2.2** Two particles impact with perpendicular initial velocities as shown. For $e = 0.5$, $\mu = -0.1$, and $m_2 = 2m_1$, determine the final velocity components at separation.

***Answers:*** $V_{1n} = 10.0$ ft/s, $V_{1t} = 11.0$ ft/s, $V_{2n} = 5$ ft/s, $V_{2t} = 0.5$ ft/s.

**2.3** A direct, collinear impact occurs between two particles with $m_1 = 0.30$ kg, $m_2 = 0.12$ kg, and $e = 0.8$. The final velocities are $V_{1n} = 5$ m/s and $V_{2n} = 8$ m/s. Determine the initial velocities.

***Answers:*** $v_{1n} = 6.93$ m/s, $v_{2n} = 3.18$ m/s.

**2.4** Plot energy loss $T_L$ as a function of impulse ratio $\mu$ for the data corresponding to problem 2.2 for $\mu = 0$ and $\mu = -1.2$.

**2.5** Show that with external impulses $P_{1n}$ and $P_{2n}$ applied to masses $m_1$ and $m_2$, respectively, Eq. 2.23.1 becomes

$$eP_n + e\left(\frac{\overline{m}}{m_1} P_{1n} - \frac{\overline{m}}{m_2} P_{2n}\right) = \overline{m}(1 + e)(V_{2n} - V_{1n})$$

## REFERENCES

1. Goldsmith, W., *Impact*, Edw. Arnold, London, 1960.
2. Brach, R. M., ''Impact Dynamics with Applications to Solid Particle Erosion,'' *Int. J. of Impact Eng.*, 7(1), 37-54, 1988.
3. Carden, H. D., ''Impact Analysis of Airplane Crash Data with Consideration Given to Human Tolerance,'' Paper No. 830748, SAE, Warrendale, PA. 1983
4. Maw, N., Barber, J. R., and Fawcett, I. N., ''The Role of Elastic Tangential Compliance in Oblique Impact,'' *Trans. ASME J. Lub. Tech.*, **103,** 74, 1981.
5. Grant, G. and Tabakoff, W., ''Erosion Prediction in Turbomachinery Resulting from Environmental Solid Particles,'' *J. Aircraft*, **12,** 471–478, 1975.

# CHAPTER 3

# RESTITUTION, FRICTION, AND ENERGY LOSS

## 3.1 INTRODUCTION

The energy loss associated with an impact can occur in many forms. Energy can be transferred or lost as elastic strain energy, plastic strain energy, fracture, light (sparks and triboluminescence), sound, and so forth. At high enough velocities, material temperatures can rise to levels where localized melting and flow can occur. For the purposes of the dynamics as considered here, the final disposition of the kinetic energy that is lost is not particularly significant. Even very-low-speed collisions with perfectly elastic deformations can lose energy, since initial rigid body kinetic energy can be converted to internal vibrations and waves that persist after contact ends. This aspect of the impact process is not covered here. Several excellent sources of information and methods are available.[1-3] What is covered here in detail is the effect of energy loss on the rigid body motion.

Energy-conservation principles of mechanics apply to collisions as to any other dynamics problem. In the absence of active input from the environment, one does not expect the kinetic energy to increase as a result of an impact of a system of particles or rigid bodies. With mechanisms such as friction and elastic/plastic deformation in the region of contact, system energy losses are expected in a realistic collision. In the approach developed in this book, the energy losses are controlled by the coefficients. This chapter will provide the relationships between kinetic energy, these coefficients, and the velocity changes.

So far, three coefficients have been used. These are the coefficient of restitution $e$, the impulse ratio $\mu$, and the tangential restitution coefficient $e_t$. The last two were shown to be directly related (see Eq. 2.29) and so the discussion here will refer to $e$ and $\mu$ as independent constants when dealing with impact dynamics. For a given collision, each is a constant even though their value may depend on the initial velocities and angle of incidence for a class of collisions. Furthermore, they may not be independent from the point of view of local material behavior, but, since they appear in the system equations, they can be treated as independent constants in the manner in which they control velocity changes. Furthermore, their effects can be viewed such that $e$ controls the "normal" velocity changes and energy loss and $\mu$ controls the "tangential" velocity changes and energy loss. For point mass or particle impacts, these associations are relatively clear. But later, when rigid body rotations are permitted, the roles of $e$ and $\mu$ are less distinct.

Another characteristic of the coefficients $e$ and $\mu$ is that they play a dual role in the impact process. Equation 1.20 shows that $e^2$ is related to energy loss and is treated by many as an energy-loss parameter. The usual definition of $e$, however (see Eq. 1.15), is actually a kinematic constraint. Values of $e$ can be chosen between 0 and 1, which directly control the relative normal velocity ratios. As will be seen later in this chapter, the same dual role is played by the impulse ratio $\mu$. As $\mu$ is varied between 0 and a critical value $\mu_c$, the relative tangential velocity ratio also changes. Lack of recognition of this dual role sometimes causes incorrect or at least unrealistic solutions of impact problems. It is generally recognized that $e$ is limited to values between 0 and 1 (although, as seen earlier, negative values do have a realistic physical meaning). The existence and properties of the critical or limiting value of $\mu$ is less well known. Further confusion over the distinction between $\mu$ and the Coulomb friction coefficient $f$ makes matters even worse. Another, related concept enters into the role of $\mu$. That is the question of whether or not relative tangential motion (sliding) still exists at the time of separation. The view taken here is that an impact is a dynamic event and that as the two bodies just begin to touch, a relative tangential velocity will exist (in general). Whether or not this sliding continues throughout the contact duration or ends prior to separation needs to be determined. When two point masses collide with an initial relative tangential velocity of zero, the final relative tangential velocity will also be zero. That is, a central collision must remain central. This is true as long as no active tangential force mechanisms are considered that can increase relative tangential velocities (such as a local explosive charge). Later it will be seen that for collisions of rigid bodies rotational kinematics permits relative velocity changes of either sign at the contact area. Relative tangential velocity can increase and even reverse. This leads to another question when Coulomb friction supplies the tangential force in the contact region. If the relative contact velocity can reverse, it must pass through zero. At zero relative velocity, the classical Coulomb model requires the use of a static coefficient of friction. Current understanding of the mechanics of friction points out that the higher value of the friction force normally associated with static coefficient requires a finite time to be developed.[9] Consequently, for applications to impact where contact durations are short, dynamic values of the friction coefficient ordinarily

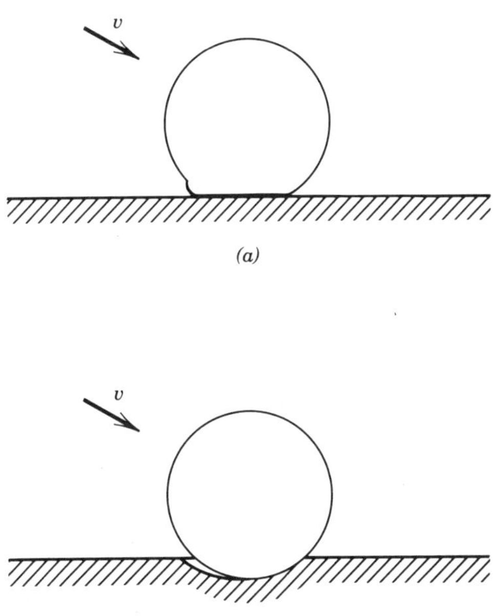


**Figure 3.1.** Extreme scenarios of relative hardness; (*a*) soft projectile, hard surface; (*b*) hard projectile, soft surface.

should be used in those circumstances when a Coulomb friction model is appropriate.

All of these concepts are intimately related to energy loss and are discussed in this chapter for point-mass collisions. In addition, the relationship between the coefficient of restitution and local material behavior is explored through some simple examples.

One final point is made before studying energy loss. In engineering, it is frequently fruitful to consider extreme cases in order to visualize trends. Consider Fig. 3.1, which depicts a soft object colliding with a hard surface in Fig. 3.1*a* and the opposite extreme in Fig. 3.1*b*. The soft object could be a loose wheel from a damaged racing vehicle, for example, colliding with the pavement. Here one might expect Coulomb friction to provide a reasonably good model for the tangential effects. A collision of a hard particle (such as steel sphere) into a soft target (such as wood) causes a significant amount of penetration. Here a simple friction model of the tangential effects would not be expected to be valid. In all cases, however, the definition of $\mu$ as an impulse ratio is valid.

## 3.2 WORK OF IMPULSES

It is often necessary to calculate the work done by impulses, since this work is directly related to changes in kinetic energy. Power is equal to the product of force and velocity, and work is the time integral of power, so the work done can be written as

$$W_r = \int_{\tau_1}^{\tau_2} F_r(\tau) v_r(\tau)\, d\tau$$

$F_r$ is a force component and $v_r$ is the velocity component along the same coordinate direction. Using Eq. 1.1 and invoking the mean value theorem from calculus permits an average value of $v_r(\tau)$ to be placed in front of the integral, giving

$$W_r = \overline{v}_r \int_{\tau_1}^{\tau_2} F_r(\tau)\, d\tau = \overline{v}_r P_r \tag{3.1}$$

where $\overline{v}_r$ is the average velocity component at the point of application of the impulse component in the direction of the impulse. The quantity $P_r$ is the component of the impulse corresponding to $F_r$.

■ *Example 3.1*

PROBLEM: Two particles with arbitrary initial velocities collide with no friction, as discussed in Chapter 1. Show that the work done by the normal impulse on the two particles is equal but opposite in sign to the kinetic energy loss, Eq. 1.20.

*Solution:* According to Eq. 3.1, the work done is the product of average velocity and the impulse $P_n$; from Eq. 2.23.2, $P_n$ is

$$P_n = (1 + e)\overline{m}(v_{2n} - v_{1n})$$

The average velocity of the point of application of $P_n$ is the relative normal velocity of the two particles. This is,

$$\overline{v}_n = \tfrac{1}{2}[(V_{1n} - V_{2n}) + (v_{1n} - v_{2n})]$$

Use of Eq. 1.13 gives

$$\overline{v}_n = \tfrac{1}{2}(1 - e)(v_{1n} - v_{2n})$$

The product of these two is the work done and is

$$W_n = -\tfrac{1}{2}\overline{m}(1 - e^2)(v_{2n} - v_{1n})^2$$

Comparison with Eq. 1.18 shows that $T_L = -W_n$. ■

## 3.3 ENERGY LOSSES DURING IMPACT: FRICTION AND RESTITUTION

For the particle impact of Chapter 2, conservation of energy requires that the kinetic energy loss must equal the negative of the work done by the two impulse components $P_n$ and $P_t$. That is, $T_L + W = 0$. Using Eq. 3.1 gives

$$T_L = -(\overline{v}_n P_n + \overline{v}_t P_t) \tag{3.2}$$

where

$$\bar{v}_n = [(V_{1n} - V_{2n}) + (v_{1n} - v_{2n})]/2 \tag{3.3}$$

and

$$\bar{v}_t = [(V_{1t} - V_{2t}) + (v_{1t} - v_{2t})]/2 \tag{3.4}$$

With the elimination of $P_n$ and $P_t$ and the use of Eqs. 2.12–2.15, the energy loss is

$$T_L = \tfrac{1}{2}\bar{m}(v_{2n} - v_{1n})^2(1 + e)[(1 - e) + 2\mu r - (1 + e)\mu^2] \tag{3.5}$$

where

$$r = (v_{2t} - v_{1t})/(v_{2n} - v_{1n}) \tag{3.6}$$

The quantity $r$ is a dimensionless initial velocity ratio related to the angle of approach of the particles. Note that for a frictionless impact, $\mu = 0$, the energy loss is

$$T_L(\mu = 0) = \tfrac{1}{2}\bar{m}(v_{2n} - v_{1n})^2(1 - e^2) \tag{3.7}$$

which is the equation most commonly found in the literature for the energy loss of a collision.

A more convenient, normalized form of the energy-loss equation, Eq. 3.5, is

$$\begin{aligned} T_L^* &= \frac{2T_L}{\bar{m}[(v_{2n} - v_{1n})^2 + (v_{2t} - v_{1t})^2]} \\ &= (1 + e)[(1 - e) + 2\mu r - (1 + e)\mu^2]/(1 + r^2) \end{aligned} \tag{3.8}$$

The denominator of $T_L^*$ will never be zero, since a collision does not occur if $v_{2n} = v_{1n}$ and $v_{2t} = v_{1t}$. Note some specical cases. For $e = 1$ and $\mu = 0$, the case of a perfectly elastic and frictionless collision, $T_L^* = 0$. For $e = 0$ and $\mu = 0$, a perfectly inelastic but frictionless impact, $T_L^* = 1/(1 + r^2)$. Contrary to what is often stated, the final system energy in this latter case cannot be zero.

## 3.4 MAXIMUM ENERGY LOSS: CRITICAL IMPULSE RATIO

As just discussed, under ideal conditions ($e = 1$ and $\mu = 0$) the energy loss in a collision is zero. This can be considered as a minimum energy loss for an impact. A maximum energy loss also exists for a collision between point masses. In order to determine this maximum, it is necessary to explore in detail the role of $\mu$, the impulse ratio.

As defined by Eq. 2.9, $\mu$ is the ratio of impulse components and can take on either negative or positive values. From Eq. 2.9, it is seen that

$$\text{sgn}(\mu) = \text{sgn}(r) \tag{3.9}$$

where $r$ is given in Eq. 3.6. The use of an incorrect sign for $\mu$ in the solution equations, 2.12–2.15, will produce incorrect results. In fact, it can be seen from Eq. 3.8 and 3.9 that the $2\mu r$ term in the energy-loss expression must always remain positive. An incorrect sign for $\mu$ will frequently lead to a negative energy "loss," clearly an indication that something is incorrect or at least unrealistic. Negative energy losses can occur in another, more subtle way, as is now discussed.

Equation 3.8 for $T_L^*$ and its functional relationship to $\mu$ provides insight into the role of this coefficient in modeling the tangential velocity changes. Figure 3.2 shows typical plots of $T_L^*(\mu)$, arbitrary relative velocity ratios $r$, and various coefficients of restitution. It illustrates the quadratic dependence of $T_L^*$ on $\mu$. Assuming temporarily that $\mu$ is a coefficient of friction. It is commonly known that the combination of friction and relative sliding dissipates energy. Intuitively, an increase in a coefficient of friction for given initial conditions (given $r$) should result in a larger energy loss. Examination of Fig. 3.2 shows that for $0 < e < 1$ this is true, but only up to a point, since these curves contain a maximum. The value of $\mu$ at a maximum, $\mu_m$, can be found by setting to zero the derivative of $T_L^*$ with respect to $\mu$. This gives

$$\mu_m = r/(1 + e) \tag{3.10}$$

Now reconsider the two sets of solution equations given in Chapter 2. The first set, Eqs. 2.12–2.15, corresponds to a sliding solution. The second set, Eqs. 2.12, 2.13, 2.16, and 2.17, corresponds to the case

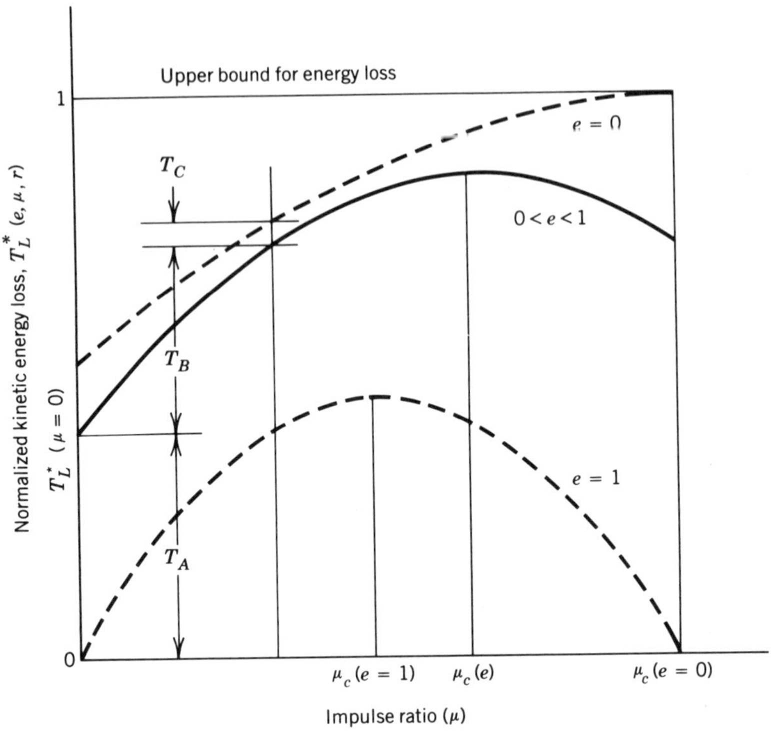


**Figure 3.2.** Variation of energy loss with impulse ratio.

when the relative tangential velocity is zero when $m_1$ and $m_2$ separate. For the special case when both solutions are identical, a value of $\mu$ is found such that

$$\mu_0 = r/(1 + e)$$

which is identical to Eq. 3.10. As the impulse ratio $\mu$ is increased in the point-mass solution equations, the energy loss increases for $\mu \leq \mu_c$. At the point $\mu = \mu_0 = \mu_m = \mu_c$, the energy loss is a maximum and the solution corresponds to zero relative tangential velocity at separation ($V_{1t} = V_{2t}$). The value $\mu_c$ will be referred to as the critical value of $\mu$.

Equations 2.14, 2.15, and 3.10 can be used to show that

$$V_{2t} - V_{1t} = (1 + e)\left(1 - \frac{\mu}{\mu_c}\right)(v_{2t} - v_{1t}) \tag{3.11}$$

As $\mu/\mu_c$ approaches 1, the final relative velocity approaches zero and the energy loss approaches its maximum. A solution for $\mu/\mu_c > 1$ corresponds to a relative tangential velocity reversal and demands that energy already removed from the system be returned (see Example 1.1). Consequently, a value of $\mu/\mu_c > 1$ corresponds to a violation of energy conservation principles for the type of tangential force mechanisms normally encountered in collisions, such as friction. On the other hand, if tangential restitution occurs, the relative tangential velocity reverses and kinetic energy is returned to the particles. An energy storage process must exist for tangential restitution to occur.

Since $\mu = \mu_c$ corresponds to $V_{1t} = V_{2t}$, there is no need for two sets of solutions to the system equations. Equations 2.14 and 2.15, which contain the impulse ratio $\mu$, are general enough to cover all planar particle impacts.

Figure 3.2 shows how energy loss can be partitioned into a portion $T_A$ associated with normal effects (restitution, $e$) and a portion $T_B$ associated with tangential effects ($\mu$). A third part $T_c$ is that stored elastically during a collision but returned during rebound.

An alternative demonstration of the role of $\mu_c$ is to find the energy loss $T_L$ for arbitrary $\mu$ and equate it to $T_L$ for $V_{1t} = V_{2t}$. Using Eqs. 2.12–2.15 for the final velocities with sliding gives an energy loss,

$$
\begin{aligned}
T_L &= \tfrac{1}{2}m_1 v_{1n}^2 + \tfrac{1}{2}m_2 v_{2n}^2 - \tfrac{1}{2}m_1 V_{1n}^2 - \tfrac{1}{2}m_2 V_{2n}^2 \\
&= -\tfrac{1}{2}m_1 \left[+2\mu(\overline{m}/m_1)(1+e)v_{1t}\Delta v_n \right. \\
&\quad \left. + \mu^2(\overline{m}/m_1)^2(1+e)^2\Delta v_n^2\right] - \tfrac{1}{2}m_2\left[-2\mu(\overline{m}/m_2)\right. \\
&\quad \left. \cdot (1+e)v_{2t}\Delta v_n + \mu^2(\overline{m}/m_2)^2(1+e)^2\Delta v_n^2\right] \qquad (3.12)
\end{aligned}
$$

where $\Delta v_n = v_{2n} - v_{1n}$. With Eq. 2.16 and 2.17 (no sliding at separation) used for $V_{1t}$ and $V_{2t}$,

$$
\begin{aligned}
T_L &= \tfrac{1}{2}m_1 v_{1n}^2 + \tfrac{1}{2}m_2 v_{2n}^2 - \tfrac{1}{2}m_1 V_{1n}^2 - \tfrac{1}{2}m_2 V_{2n}^2 \\
&= -\tfrac{1}{2}m_1\left[2(\overline{m}/m_1)\, v_{1t}\Delta v_t + (\overline{m}/m_1)^2\Delta v_t^2\right] \\
&\quad -\tfrac{1}{2}m_2\left[-2(\overline{m}/m_2)v_{2t}\Delta v_t + (\overline{m}/m_2)^2\Delta v_t^2\right] \qquad (3.13)
\end{aligned}
$$

where $\Delta v_t = v_{2t} - v_{1t}$. Equating these and simplifying gives

$$\mu = r/(1 + e) = \mu_c \tag{3.14}$$

It is now clear that a limiting or critical value $\mu_c$ of the impulse ratio coefficient exists for point-mass collisions. This limiting value corresponds to a maximum energy loss; note that it is independent of the masses. For given initial conditions and for $\mu < \mu_c$, sliding takes place throughout the entire collision. For $\mu = \mu_c$ the final relative velocity of the particles is zero and corresponds to the case where sliding ends prior to or at the time of separation.

It is a worthwhile exercise to examine the behavior of the tangential force and its impulse as $\mu$ increases. For simplicity, assume that sliding is controlled by a Coulomb friction model. Figure 3.3 shows a typical normal force $F_n(\tau)$ and its impulse $P_n$. For a relatively small friction coefficient, $f < \mu_c$, sliding continues through the impact. For the same initial conditions and a larger value of $f$, say $f_2$, a value of $P_t$ develops, just sufficiently large enough to cause sliding to cease at the time of separation, $\tau_2$. For larger values of $f > f_2$, the impulse ratio is still $\mu_c$; sliding simply ends earlier during the duration of contact as indicated in the last case in Fig. 3.3.

As a result of the properties of the coefficients $e$ and $\mu$, a maximum loss of energy exists for any collision of two particles. Substitution of $\mu_c$ from Eq. 3.10 into the expression for energy loss, Eq. 3.5, gives

$$(T_L)_{\max} = \tfrac{1}{2}\overline{m}(v_{2n} - v_{1n})^2(1 - e^2 + r^2) \tag{3.15}$$

Using the definition of $r$ (Eq. 3.6), this becomes

$$(T_L)_{\max} = \tfrac{1}{2}\overline{m}\left[(v_{2t} - v_{1t})^2 + (1 - e^2)(v_{2n} - v_{1n})^2\right] \tag{3.16}$$

This expression is independent of the impulse ratio coefficient and, consequently, independent of whatever physical process exists through which the tangential force is transmitted between the masses during impact. Equation 3.16 provides a simple expression for an upper bound on the energy loss. Setting the coefficient of restitution to zero in Eq. 3.16 provides an absolute upper bound to the kinetic energy loss for any par-

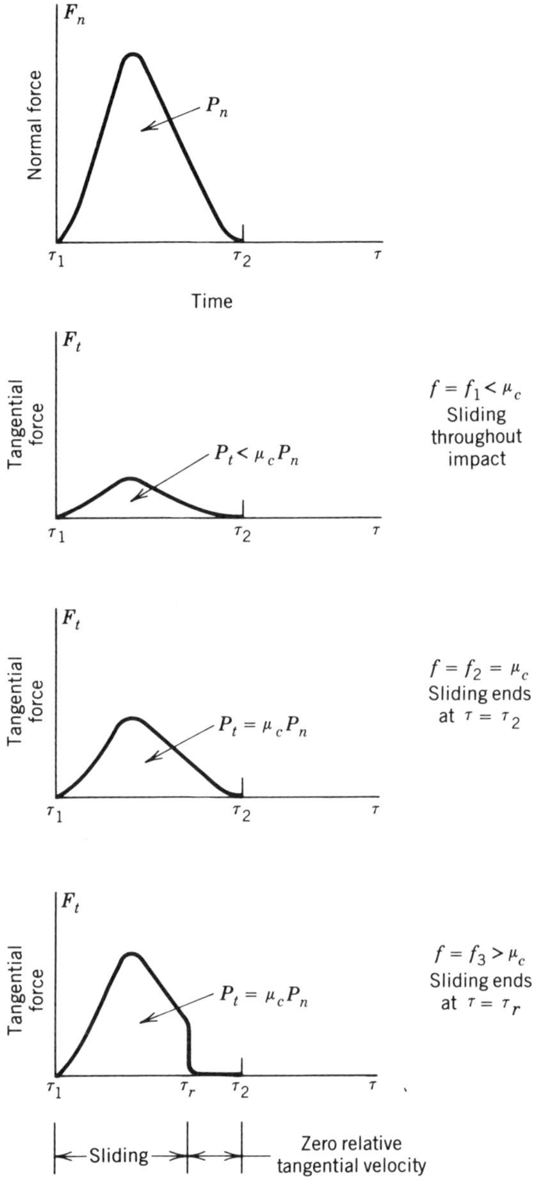


**Figure 3.3.** Transition to the critical impulse ratio.

ticle collision. $T_L$ from Eq. 3.16 with $e = 0$ is the energy loss used to normalize $T_L$ to obtain Eq. 3.8.

## ■ *Example 3.2*

PROBLEM: Two point masses move with speeds $v_1 = 12$ m/s and $v_2 = 10$ m/s in the directions shown, where $\alpha = 53.13°$. The origin of the $n$–$t$ coordinate system is placed at the point where the masses begin contact. The surfaces are rough with a dynamic friction coefficient $f = 0.6$; the restitution coefficient has a value $e = \frac{1}{3}$ and $m_2 = \frac{3}{2}m_1$.

Determine the final velocities, the maximum possible energy loss, and the actual energy loss.

*Solution:* The initial velocity ratio parameter $r$ is

$$r = \frac{v_{2t} - v_{1t}}{v_{2n} - v_{1n}} = \frac{8 - 12}{6 - 0} = -\frac{2}{3}$$

The critical impulse ratio $\mu_c$ is

$$\mu_c = \frac{r}{1 + e} = \frac{-\frac{2}{3}}{1 + \frac{1}{3}} = -\frac{1}{2}$$

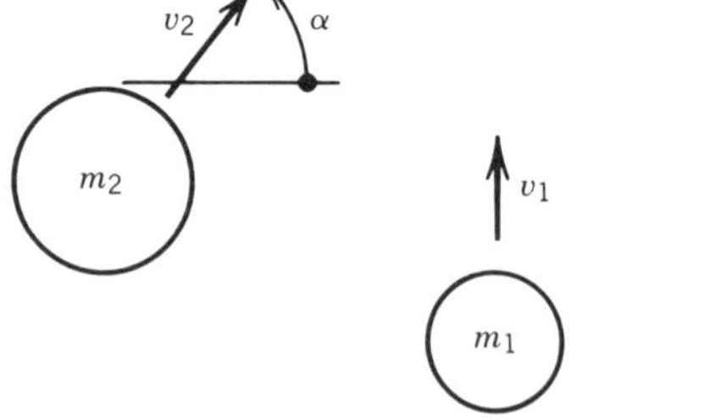


**Figure E3.2.** Particle configuration.

The negative sign on $\mu_c$ indicates that the tangential impulse $P_t$ is negative. Since $f > |\mu_c|$, sliding ends prior to separation and the solution is found using $\mu = \mu_c$. The final velocity components are found using Eqs. 2.12–2.15. These give

$$V_{1n} = 4.8 \text{ m/s}, \qquad V_{2n} = 2.8 \text{ m/s}$$

$$V_{1t} = 9.6 \text{ m/s}, \qquad V_{2t} = 9.6 \text{ m/s}$$

Following the collision, particle 1 moves at an angle of 63.4° relative to the $n$ axis. Particle 2 moves off at a corresponding angle of 73.7°. The actual energy loss can be calculated using Eq. 3.14, since $\mu = \mu_c$. This gives

$$T_L = \tfrac{1}{2}\overline{m}(v_{2n} - v_{1n})^2(1 - e^2 + r^2) = 14.4m_1, \qquad \text{J}$$

The maximum energy loss comes from the same equation with $e = 0$. This is

$$(T_L)_{\max} = 15.6m_1, \qquad \text{J}$$

■

## 3.5 DEFORMATION AND THE COEFFICIENT OF RESTITUTION

The impact model treated previously and throughout the rest of this book deals with collision dynamics from a "before-and-after" perspective. Typically, the initial velocities and the system are given and the final velocities and energy loss are sought. Practically nothing is learned about what happens during the impact. Only general concepts of deformation are assumed, such as compression and shear in the contact regions of the masses and some elastic restoration during rebound. Modeling of the actual dynamic behavior of the masses during the impact can be quite complex. Linear material models are frequently inappropriate; analytical models are frequently insufficient. However, models of local deformation can yield a great deal of dynamic information concerning variables such as vibrational displacements, stresses, flow, and onset of fracture. Another result of deformation models (either analytical or numerical) is that they can provide values of $e$ and $\mu$ for use in the equations presented here. Of course, a commonly used alternative to deformation models is the use of experiments to provide measured values of the coefficients.

Some simple analytical models that relate to the coefficient of restitution $e$ are reviewed here. Coulomb friction is the most commonly used analytical model for tangential forces. Only a few others are known to exist, and so analytical models for estimating values of $\mu$ are not discussed here. Some of these models[4,5] will be discussed subsequently.

### 3.5.1 Restitution and Mass-Center Kinematics

Before examining particular examples, it is worthwhile to develop a reasonably general expression for the coefficient of restitution in terms of the particles' relative mass-center motion. Consider two deformable particles in contact during impact as shown in Fig. 3.4. These have a com-

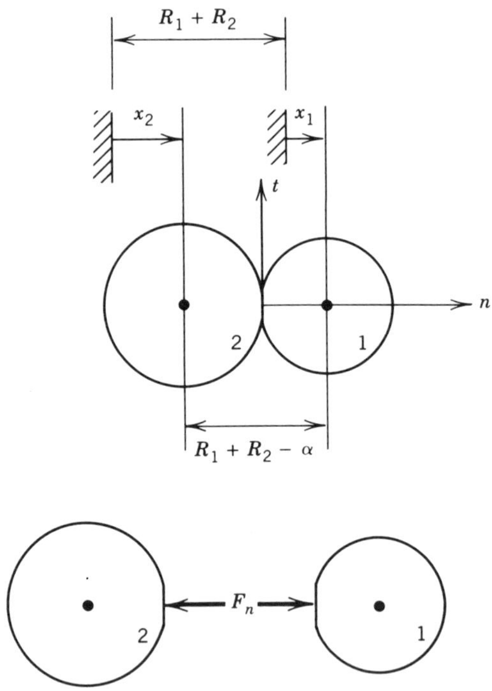


**Figure 3.4.** Coordinates and free body diagrams of colliding particles.

mon resultant normal force $F_n$ at the area of contact due to compressive surface stress $q$. If the mass centers have displacements $x_1(\tau)$ and $x_2(\tau)$, then

$$m_1\ddot{x}_1 = +F_n \tag{3.17}$$

and

$$m_2\ddot{x}_2 = -F_n \tag{3.18}$$

Multiplication of Eq. 3.17 by $m_2/(m_1 + m_2)$ and Eq. 3.18 by $m_1/(m_1 + m_2)$ and subtracting gives

$$\overline{m}\ddot{x}_1 - \overline{m}\ddot{x}_2 = F_n \tag{3.19}$$

If a relative motion coordinate is now defined, such that $\alpha(\tau) = x_2(\tau) - x_1(\tau)$, Eq. 3.19 becomes

$$\overline{m}\ddot{\alpha} + F(\dot{\alpha}, \alpha, \tau) = 0 \tag{3.20}$$

where $F(\dot{\alpha}, \alpha, \tau) = F_n$. Multiplication by $\dot{\alpha}$ and integrating over the contact duration $\tau_2 - \tau_1$ gives

$$\tfrac{1}{2}\overline{m}\dot{\alpha}_2^2 - \tfrac{1}{2}\overline{m}\dot{\alpha}_1^2 + \int F(\dot{\alpha}, \alpha, \tau)\, d\alpha = 0 \tag{3.21}$$

The first two terms represent the change in kinetic energy of the system. To show this, note that by previous notation, $\dot{\alpha}_2 = V_{2n} - V_{1n}$ and $\dot{\alpha}_1 = v_{2n} - v_{1n}$. Using Eq. 2.6 for $\mu = 0$, Eq. 3.21 becomes

$$-\tfrac{1}{2}\overline{m}(1 - e^2)(v_{2n} - v_{1n})^2 + \int F(\dot{\alpha}, \alpha, t)\, d\alpha = 0 \tag{3.22}$$

The first term is recognized as the energy loss (see Eq. 3.7), so that

$$T_L = \int F(\dot{\alpha}, \alpha, \tau)\, d\alpha \tag{3.23}$$

This shows that the coefficient of restitution is directly related to the energy lost by normal deformation of the particles during the collision.

That is,

$$e^2 = 1 - \frac{2}{\overline{m}(v_{2n} - v_{1n})^2} \int F(\dot{\alpha}, \alpha, \tau)\, d\alpha \tag{3.24}$$

If the intercollision force can be broken down into position- and rate-dependent parts and does not depend explicitly on time $\tau$, then

$$F(\dot{\alpha}, \alpha, \tau) = F_{\dot{\alpha}}(\dot{\alpha}) + F_{\alpha}(\alpha) \tag{3.25}$$

and

$$\int F(\dot{\alpha}, \alpha, \tau)\, d\alpha = \int_{\tau 1}^{\tau 2} \dot{\alpha} F_{\dot{\alpha}}\, d\tau + \int_{\alpha 1}^{\alpha 2} F_{\alpha}\, d\alpha \tag{3.26}$$

The impact duration can be estimated using

$$\tau = \tau_2 - \tau_1 = \int_{\alpha 1}^{\overline{\alpha}} \frac{1}{\dot{\alpha}}\, d\alpha + \int_{\overline{\alpha}}^{\alpha 2} \frac{1}{\dot{\alpha}}\, d\alpha \tag{3.27}$$

where $\overline{\alpha}$ is the displacement at maximum approach.

Goldsmith[1] points out that when two elastic spheres of different materials collide, the coefficient of restitution of the combination, $e_{12}$, has been observed to be related to the corresponding like-material coefficients, $e_{11}$ and $e_{22}$, by the formula

$$e_{12} = (e_{11}E_2 + e_{22}E_1)/(E_1 + E_2) \tag{3.28}$$

where $E_1$ and $E_2$ are the respective moduli of elasticity.

Barkan[6] provides several examples of the preceding equations. Results of two are presented here.

### 3.5.2 Linear Elastic Collision[6]

In this case, $e = 1$. For $F(\dot{\alpha}, \alpha, \tau) = k\alpha$, where $k$ is a stiffness, the peak force is

$$F_p = (v_{1n} - v_{2n})\sqrt{\overline{m}k} \tag{3.29}$$

and the duration of impact is

$$\tau = \pi\sqrt{\overline{m}/k} \tag{3.30}$$

The corresponding case of an elastic impact with a massive, rigid surface can be examined by letting one of the masses grow indefinitely large.

### 3.5.3 Linear Plastic Collision[6]

Figure 3.5 illustrates the force deformation relationship for an idealized linear, perfectly plastic material. In this case,

$$e = -\frac{F_p}{\dot{\alpha}_1}\sqrt{\frac{1}{\overline{m}k}} \tag{3.31}$$

Equation 3.31 can furnish values of $e > 1$. If so, no plastic deformation takes place and $e = 1$. The impact duration is

$$\tau = \frac{\overline{m}\dot{\alpha}_1}{F_P}\left[1 - \left(\frac{F_P\eta}{k\dot{\alpha}_1}\right)^2\right] + \frac{\pi}{2\eta} + \frac{1}{\eta}\sin^{-1}\left(\frac{F_P\eta}{k\dot{\alpha}_1}\right), \qquad \eta = \sqrt{k/\overline{m}} \tag{3.32}$$

### 3.5.4 Hertz Theory of Impact

The geometry considered here is as illustrated in Fig. 3.4, where two spheres are shown for simplicity. Other particle shapes are permitted as

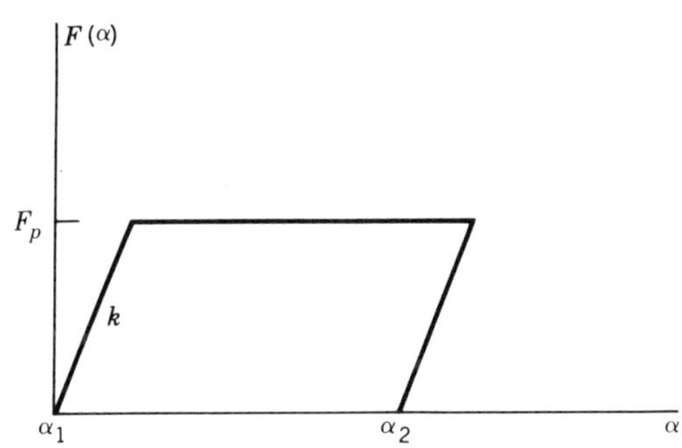


**Figure 3.5.** Idealized linear plastic force.

long as the contact regions are near spherical and the impact is central. (In fact, a rod with a hemispherically shaped end is often used in impact experiments.[1]) Only normal, collinear motion is considered. According to Hertz as developed by Timoshenko and Goodier[7]

$$\alpha = q_0 \pi^2 a (k_1 + k_2)/2 \tag{3.33}$$

and

$$a = q_0 \pi^2 \bar{r} (k_1 + k_2)/2 \tag{3.34}$$

The quantity $a$ is the radius of the contact circle, $q_0$ is pressure, and

$$\begin{aligned} \bar{r} &= r_1 r_2/(r_1 + r_2) \\ k_i &= (1 - \nu_i^2)/\pi E_i, \qquad i = 1, 2 \end{aligned} \tag{3.35}$$

where $\nu_i$ and $E_i$ are Poisson's ratio and Young's modulus, respectively, for each particle. A hemispherical pressure distribution between the particles is assumed, which has the maximum value of $q_0$. Under static conditions and using these equations it is possible to show that the normal force is related to the relative displacement by

$$F_n(\alpha) = k\alpha^{3/2} \tag{3.36}$$

where $k = 4\sqrt{\bar{r}}/3\pi(k_1 + k_2)$. This is an elastic model and the energy loss will be zero and the value of $e$ will be 1. The relative velocity at any time during contact is

$$\dot{\alpha}^2 = \dot{\alpha}_1^2 - \frac{2}{\bar{m}} \int_0^{\alpha} k\alpha^{3/2}\, d\alpha = \dot{\alpha}_1^2 - \frac{4}{5} \frac{k\alpha^{5/2}}{\bar{m}} \tag{3.37}$$

Equation 3.37 can provide the maximum displacement by setting $\dot{\alpha}$ to zero. The duration of impact is

$$\tau = 3.214\, (\bar{m}/k)^{2/5} \dot{\alpha}_1^{-1/5} \tag{3.38}$$

If a collision is inelastic, the preceding is not directly applicable. To synthetically introduce energy loss, it is possible to assume that Eq. 3.36 is valid for the initial approach and rebound during impact. The approach also includes a perfectly plastic deformation phase with no rate dependence from $\alpha_p$ to $\alpha_m$. This is shown schematically in Fig. 3.6. Consequently, deformation is elastic during approach of the mass centers for $\alpha$ less than $\alpha_p$ at which perfectly plastic deformation begins. During rebound, for $\alpha$ less than $\alpha_m$ ($\alpha_m$ is the maximum relative displacement), deformation is again elastic. From Eq. 3.24, this gives

$$e^2 = 1 - \frac{2k\alpha_p^{3/2}(\alpha_m - \alpha_p)}{\overline{m}\dot{\alpha}_1^2} \tag{3.39}$$

After extensive review of the above Hertzian theory, Goldsmith[1] concludes that the Hertz theory provides a good model of the collision of two spheres or the impact of a sphere against a thick plate only if the materials are hard and the initial velocity is low. For softer materials or higher impact velocities, plastic deformation and strain rate effects must be included.

### 3.5.5 Impact With Viscous Damping

Figure 3.7 shows a physical model occasionally used as an engineering impact model where energy is dissipated through viscous effects. It is claimed that a linear viscous model is not realistic[8] because an initial velocity causes a discontinuous viscous force in a linear viscous damper. This is true, of course. Nevertheless, the model is presented because it

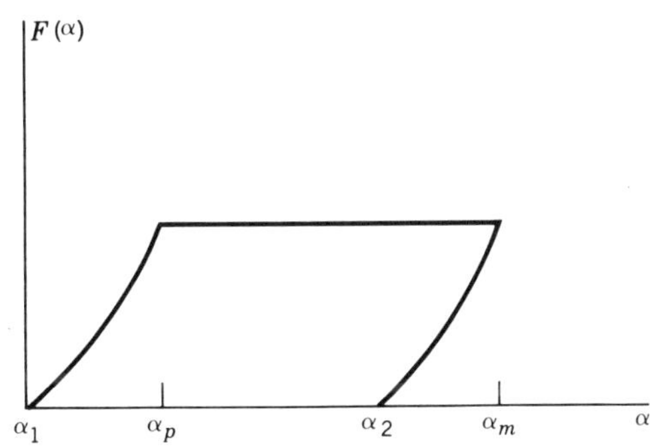


**Figure 3.6.** Idealized nonlinear plastic force.

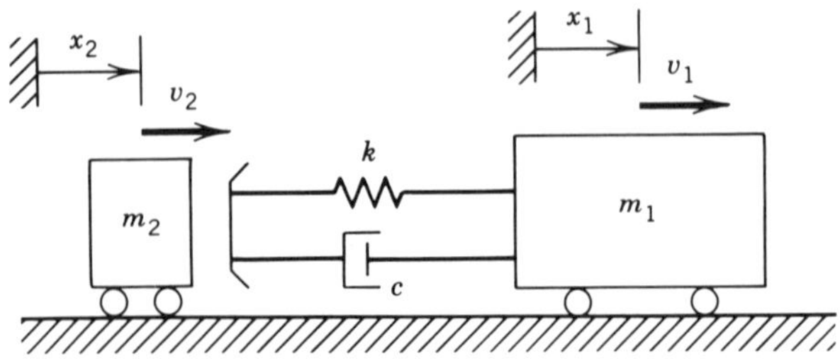


**Figure 3.7.** Linear viscous damper model.

has such widespread use in vibration analysis. After contact has initiated the equations of motion for the two masses are

$$m_1\ddot{x}_1 + c\dot{x}_1 - c\dot{x}_2 + kx_1 - kx_2 = 0 \tag{3.40.1}$$

and

$$m_2\ddot{x}_2 - c\dot{x}_1 + c\dot{x}_2 - kx_1 + kx_2 = 0 \tag{3.40.2}$$

The constant $c$ is the damping coefficient, $k$ is a linear spring rate, and $m_1$ and $m_2$ are the masses.

A change to relative coordinates greatly simplifies these equations. Let $\alpha' = (m_1x_1 + m_2x_2)/(m_1 + m_2)$ and $\alpha = x_2 - x_1$. This gives

$$\ddot{\alpha}' = 0 \tag{3.41}$$

and

$$\ddot{\alpha} + 2\zeta\omega_n\dot{\alpha} + \omega_n^2\alpha = 0 \tag{3.42}$$

where $2\zeta\omega_n = c/\overline{m}$ and $\omega_n^2 = k/\overline{m}$. This is standard notation from the field of vibration theory where $\omega_n$ is the natural frequency in units of radians per second and $\zeta$ is the damping ratio. For underdamped systems, $0 < \zeta < 1$. The solution of Eq. 3.42 is well known. For zero initial displacement and $\dot{\alpha}(0) = v_2 - v_1$,

$$\alpha(\tau) = \frac{v_2 - v_1}{\omega_d} \exp(-\zeta\omega_n\tau) \sin \omega_d\tau \tag{3.43}$$

where $\omega_d^2 = \omega_n^2(1 - \zeta^2)$. The solution, Eq. 3.43, is periodic with period $2\pi/\omega_d$. However, when used as an impact model, the system remains in contact for one half of a cycle, so the impact duration is $\pi/\omega_d$. The final relative velocity is $V_2 - V_1 = \dot{\alpha}(\pi/\omega_d)$. This can be used to find the coefficient of restitution, which is

$$e = \exp\left(-\pi\zeta/\sqrt{1-\zeta^2}\right) \tag{3.44}$$

The coefficient $e$ depends exclusively on a single system property, the damping ratio $\zeta$. It does not depend on the initial velocities. Note that for no damping ($\zeta = 0$) the collision is entirely elastic, $e = 1$. For $\zeta = 1$ and $c = c_{cr}$, commonly called critical damping, $e = 0$. This occurs for $c = c_{cr} = 2\sqrt{k\overline{m}}$.

## 3.6 NONLINEAR OSCILLATOR MODEL

Hunt and Crossley[10] present an impact model based on a mass–spring–damper system as previously discussed, but with the spring and damper both nonlinear. Their equation of motion is

$$m\ddot{\alpha} + \beta k \alpha^p \dot{\alpha} + k\alpha^p = 0 \tag{3.45}$$

where $\beta$ is a damping parameter, $k$ is a spring parameter, and $p$ is an arbitrary exponent. The form of the velocity term with $\alpha^p$ as a factor

**Figure 3.8.** Hysteresis-type impact force of Hunt and Crossley.[10]

removes the force discontinuity at the initiation of impact. The force–displacement curve takes on the form of a hysteresis curve as shown in Fig. 3.8. This model can serve well for single impacts as well as vibrational impacts (see Chapter 9). Because Eq. 3.45 is nonlinear, its solution must be found numerically, however. It is shown[10] that for small $\beta$ and low velocity impact, the coefficient of restitution is given approximately by

$$e = 1 - \beta v_n \tag{3.46}$$

## 3.7 SUMMARY OF EQUATIONS FOR A TWO-PARTICLE COLLISION

Equations 2.12–2.15 are the solution equations for the final velocities following a planar collision of two point masses. The form of the tangential velocities can be made more convenient by eliminating $1 + e$ and introducing $\mu_c$ through the use of Eq. 3.10. This gives

$$V_{1n} = v_{1n} + \frac{\overline{m}}{m_1}(1 + e)(v_{2n} - v_{1n}) \tag{3.47}$$

$$V_{1t} = v_{1t} + \frac{\overline{m}}{m_1}\frac{\mu}{\mu_c}(v_{2t} - v_{1t}), \; |\mu| \leq |\mu_c| \tag{3.48}$$

$$V_{2n} = v_{2n} - \frac{\overline{m}}{m_2}(1 + e)(v_{2n} - v_{1n}) \tag{3.49}$$

$$V_{2t} = v_{2t} - \frac{\overline{m}}{m_2}\frac{\mu}{\mu_c}(v_{2t} - v_{1t}), \; |\mu| \leq |\mu_c| \tag{3.50}$$

where

$$\mu_c = \frac{r}{1 + e} \tag{3.10}$$

and

$$r = \frac{v_{2t} - v_{1t}}{v_{2n} - v_{1n}} \tag{3.6}$$

The kinetic energy loss of the collision $T_L$ can be written as

$$T_L = \tfrac{1}{2}\overline{m}(v_{2n} - v_{1n})^2(1 + e)[(1 - e) + 2\mu r - (1 + e)\mu^2] \tag{3.5}$$

This can be placed into a nondimensional, normalized form using Eq. 3.8 and the definition of a relative angle of approach, $\alpha$, such that

$$\alpha = \tan^{-1}(1/r) \tag{3.51}$$

Then the energy loss can be written in the form

$$T_L^* = (1 - e^2)\sin^2\alpha + \left[2\frac{\mu}{\mu_c} - \left(\frac{\mu}{\mu_c}\right)^2\right]\cos^2\alpha \tag{3.52}$$

The maximum energy that can be lost during a collision of two point masses is given by $T_L$ for $e = 0$ and $\mu = \mu_c$. From Eq. 3.5 this is

$$(T_L)_{\max} = \tfrac{1}{2}\overline{m}[(v_{2n} - v_{1n})^2 + (v_{2t} - v_{1t})^2] \tag{3.53}$$

## 3.8 PROBLEMS

**3.1** A cue ball, $m_1$, has coordinates $(x_1, y_1)$ and a target ball, $m_2$, has coordinates $(x_2, y_2)$, as shown. Using point-mass impact theory, determine the angle $\alpha_n$ for arbitrary $e$ and $\mu$ such that an impact drives the target ball straight toward the pocket at the origin. Show that for $\mu = 0$, $\alpha_n = \beta$.

**3.2** Let $m_2$ become indefinitely large and $m_1 = m$. Show that the solution equations for a particle striking a moving rigid surface are

$$V_n = v_n + (1 + e)(v_{sn} - v_n)$$

$$\text{and} \qquad V_t = v_t + \frac{\mu}{\mu_c}(v_{st} - v_t)$$

where $v_{sn}$ and $v_{st}$ are the velocity components of the surface.

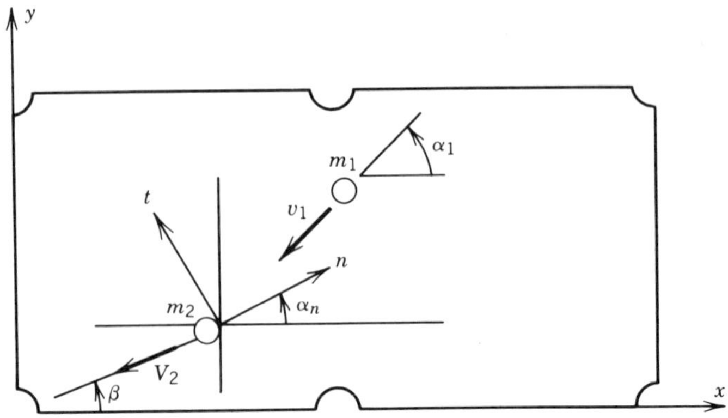


**Figure P3.1.** Billiard configuration.

**3.3** Solve Example 1.4 with the addition of sliding friction with a coefficient of $f = 0.10$. Determine the energy loss, the percentage lost due to friction, and the percentage of energy lost due to normal inelasticity. A hockey puck weighs 5.5 ounces.

***Answers:*** $T_{Lf} = 21.0\%$ and $T_{Le} = 79.0\%$.

**3.4** A spherical particle is propelled toward and strikes a flat massive surface with a speed of $v_0 = 300$ m/s and angle of incidence $\alpha$. The angle $\alpha$ is measured relative to the surface. The coefficient of restitution is given by

$$e = \frac{k}{k + v_n}$$

where $k = 100$ m/s. The impulse ratio $\mu$ is governed primarily by Coulomb friction with $f = 0.35$. Plot energy loss as a function of angle of incidence for $0 \leq \alpha \leq 90°$.

## REFERENCES

1. Goldsmith, W., *Impact*, Edw. Arnold, 1960.
2. Johnson, W., *Impact Strength of Materials*, Edw. Arnold, London 1972.
3. Zukas, J. A., *Impact Dynamics*, John Wiley, New York, 1982.

4. Hutchings, I. M., Macmillan, N. H., and Rickerby, D. G., "Further Studies of the Oblique Impact of a Hard Sphere Against a Ductile Solid," *Int. J. Mech. Sci.*, **23**(11), 639–646, 1981.
5. Maw, N., Barber, J. R., and Fawcett, J. N., "The Role of Elastic Tangential Compliance in Oblique Impact," *Trans. ASME, J. Lub. Tech.*, **103,** 74–80, 1981.
6. Barkan, P., "Impact," in *Mechanical Design and Systems Handbook*, Section 16. McGraw-Hill, New York, 1972.
7. Timoshenko, S. and Goodier, J. N., *Theory of Elasticity*, 2nd ed., McGraw-Hill, New York, 1951.
8. Herbert, R. G. and McWhannell, D. C., "Shape and Frequency Composition of Pulses From an Impact Pair," *Trans. ASME, J. Engrg. Ind.*, **99,** 513–518, 1977.
9. Suh, N. P. and Saka, N., *Fundamentals of Tribology*, Proceedings of an International Conference on the Fundamentals of Tribology, MIT Press, Cambridge, MA, June 1978.
10. Hunt, K. H. and Crossley, F. R. E., "Coefficient of Restitution Interpreted as Damping in Vibroimpact," *Trans. ASME, J. Appl. Mech.*, **97,** 440–445, 1975.

# CHAPTER 4

# THREE-DIMENSIONAL PARTICLE COLLISIONS

The equations of planar particle collisions from Chapter 2 and the concepts behind the impact process from Chapter 3 extend quite naturally to three dimensions. The analysis of planar collisions of point masses in Chapter 2 consists of four system equations and four unknown final velocity components. The normal and tangential impulses were also unknown, but they were eliminated from the system equations and computed once all velocities are known. This will be done here also, but first the system equations will be presented in a form where the unknown impulse components are retained as explicit unknowns.

A free body diagram of particle 2 and the coordinate system is illustrated in Fig. 4.1, which is a generalization of Fig. 1.2. The principle of impulse and momentum for particle one along the three axes $n$, $t$, and $t'$, are, respectively,

$$m_1 V_{1n} - m_1 v_{1n} = P_n \tag{4.1}$$

$$m_1 V_{1t} - m_1 v_{1t} = P_t \tag{4.2}$$

$$m_1 V_{1t'} - m_1 v_{1t'} = P_{t'} \tag{4.3}$$

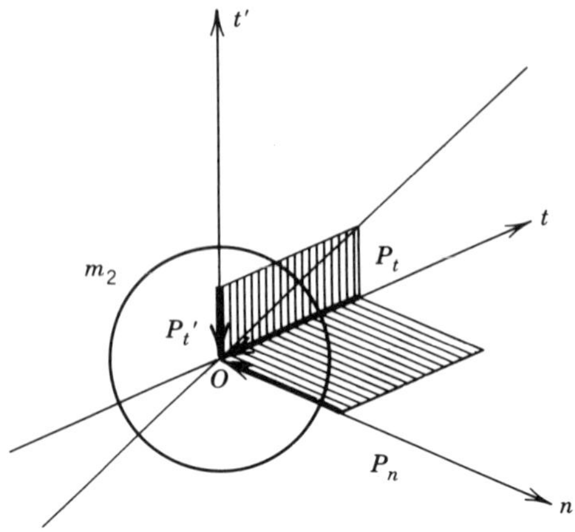


**Figure 4.1.** Three-dimensional free body diagram of mass 2 with impulse components.

For particle 2, the corresponding equations are

$$m_2 V_{2n} - m_2 v_{2n} = -P_n \tag{4.4}$$

$$m_2 V_{2t} - m_2 v_{2t} = -P_t \tag{4.5}$$

$$m_2 V_{2t'} - m_2 v_{2t'} = -P_{t'} \tag{4.6}$$

The concept of restitution normal to the $t - t'$ plane provides

$$V_{1n} - V_{2n} = -e(v_{1n} - v_{2n}) \tag{4.7}$$

Finally, two constants $\mu_t$ and $\mu_{t'}$ are defined as impulse ratios giving

$$P_t = \mu_t P_n \tag{4.8}$$

$$P_{t'} = \mu_{t'} P_n \tag{4.9}$$

The coefficients $\mu_t$ and $\mu_{t'}$ are constants. Their values are established by the mechanisms of tangential force generation in the $t$ and $t'$ directions, respectively. At this point, they may or may not be coefficients of friction, but can be referred to as equivalent coefficients of friction with no loss in generality.

The preceding equations form a set of nine linear algebraic equations that can be solved for the nine unknowns: $V_{1n}$, $V_{1t}$, $V_{1t'}$, $V_{2n}$, $V_{2t}$, $V_{2t'}$, $P_n$, $P_t$, and $P_{t'}$. Elimination of the impulse components can be done easily to reduce the problem to six equations and six unknowns. Judicious addition of pairs of equations, starting with Eqs. 4.1 and 4.4, results in the expressions of conservation of momentum along the three axes. These are

$$m_1 V_{1n} + m_2 V_{2n} = m_1 v_{1n} + m_2 v_{2n} \tag{4.10}$$

$$m_1 V_{1t} + m_2 V_{2t} = m_1 v_{1t} + m_2 v_{2t} \tag{4.11}$$

$$m_1 V_{1t'} + m_2 V_{2t'} = m_1 v_{1t'} + m_2 v_{2t'} \tag{4.12}$$

Equation 4.7 provides a fourth equation and needs no modification. The last two equations are found by eliminating $P_n$, $P_t$, and $P_{t'}$ from Eqs. 4.8 and 4.9; the result can be put into the form

$$\begin{aligned} &\mu_t m_1 V_{1n} - m_1 V_{1t} - \mu_t m_2 V_{2n} + m_2 V_{2t} \\ &\quad = \mu_t m_1 v_{1n} - m_1 v_{1t} - \mu_t m_2 v_{2n} + m_2 v_{2t} \end{aligned} \tag{4.13}$$

$$\begin{aligned} &\mu_{t'} m_1 V_{1n} - m_1 V_{1t'} - \mu_{t'} m_2 V_{2n} + m_2 V_{2t'} \\ &\quad = \mu_{t'} m_1 v_{1n} - m_1 v_{1t'} - \mu_{t'} m_2 v_{2n} + m_2 v_{2t'} \end{aligned} \tag{4.14}$$

Equations 4.10–4.14 and 4.7 are the system equations and can be solved for the unknown final velocities. Rather than do that, however, it is instructive to make one more change of equations. In problems involving momentum, it is frequently convenient to define momentum variables or, in some cases, variables representing momentum change. Define a momentum variable $U_{ij}$, where

$$U_{ij} = m_i (V_{ij} - v_{ij}), \qquad i = 1, 2 \text{ and } j = n, t, t' \tag{4.15}$$

With these new variables defined, the six system equations become

$$U_{1n} + U_{2n} = 0 \tag{4.16}$$

$$U_{1t} + U_{2t} = 0 \tag{4.17}$$

$$U_{1t'} + U_{2t'} = 0 \tag{4.18}$$

$$U_{1n}/m_1 - U_{2n}/m_2 = -(1 + e)(v_{1n} - v_{2n}) \quad (4.19)$$

$$\mu_t U_{1n} - U_{1t} - \mu_t U_{2n} + U_{2t} = 0 \quad (4.20)$$

$$\mu_{t'} U_{1n} - U_{1t'} - \mu_{t'} U_{2n} + U_{2t'} = 0 \quad (4.21)$$

In this form, the solution is relatively simple to determine by elimination. The solution equations for the three-dimensional particle collision problem are

$$U_{1n} = m_1(V_{1n} - v_{1n}) = \overline{m}(1 + e)(v_{2n} - v_{1n}) \quad (4.22)$$

$$U_{2n} = m_2(V_{2n} - v_{2n}) = -\overline{m}(1 + e)(v_{2n} - v_{1n}) \quad (4.23)$$

$$U_{1t} = m_1(V_{1t} - v_{1t}) = \mu_t\overline{m}(1 + e)(v_{2n} - v_{1n}) \quad (4.24)$$

$$U_{2t} = m_2(V_{2t} - v_{2t}) = -\mu_t\overline{m}(1 + e)(v_{2n} - v_{1n}) \quad (4.25)$$

$$U_{1t'} = m_1(V_{1t'} - v_{1t'}) = \mu_{t'}\overline{m}(1 + e)(v_{2n} - v_{1n}) \quad (4.26)$$

$$U_{2t'} = m_2(V_{2t'} - v_{2t'}) = -\mu_{t'}\overline{m}(1 + e)(v_{2n} - v_{1n}) \quad (4.27)$$

The first four solution equations are identical to those of the planar problem. With all velocity components known, the impulse components $P_n$, $P_t$, and $P_{t'}$ can be computed.

## 4.1 ENERGY LOSS

The kinetic energy loss $T_L$ can be found by subtracting the final kinetic energy from the initial kinetic energy. The final velocities are eliminated using Eqs. 4.22–4.27; this results in

$$T_L = \tfrac{1}{2}\overline{m}(v_{2n} - v_{1n})^2(1 + e)[1 - e + 2(\mu_t r + \mu_{t'} r') - (1 + e)(\mu_t^2 + \mu_{t'}^2)] \quad (4.28)$$

where

$$r = (v_{2t} - v_{1t})/(v_{2n} - v_{1n})$$

and

$$r' = (v_{2t'} - v_{1t'})/(v_{2n} - v_{1n})$$

$T_L$ reaches a global maximum relative to the impulse ratios when

$$\mu_{tc} = r/(1 + e) \tag{4.29}$$

and

$$\mu_{t'c} = r'/(1 + e) \tag{4.30}$$

The solution equations can be used to show that $V_{1t} - V_{2t} = 0$ and $V_{1t'} - V_{2t'} = 0$ occur when $\mu_t$ and $\mu_{t'}$ are at these critical values. This corresponds to the resultant tangential impulse $P_{tg}$ reaching a value of

$$P_{tg} = \sqrt{P_t^2 + P_{t'}^2} = \sqrt{\mu_t^2 + \mu_{t'}^2}\, P_n \tag{4.31}$$

If

$$\mu_t < \mu_{tc} \qquad \text{and} \qquad \mu_{t'} < \mu_{t'c}$$

then the final relative tangential velocity is nonzero. The preceding problem formulation and solution equations permit proper treatment of the rare case when the relative tangential velocity in one direction ends prior to separation, but does not in the other. When

$$\mu_t = \mu_{tc} \qquad \text{and} \qquad \mu_{t'} = \mu_{t'c}$$

then the final relative tangential velocity is zero. If the tangential force system is isotropic, which is the case typically encountered, then $\mu_t = \mu \cos \eta$ and $\mu_{t'} = \mu \sin \eta$. A single critical impulse ratio $\mu_c$ results:

$$P_{tg} = \mu_c P_n \tag{4.32}$$

In particular, if relative tangential motion is controlled by Coulomb friction with a coefficient $f$, then sliding continues throughout the collision for $f < \mu_c$. The final relative velocity will be zero for $f \geq \mu_c$.

## ■ *Example 4.1*

PROBLEM: A cylindrical object, $m_1$, moves vertically upward at a speed of 300 km/hr. A small sphere with mass $m_2$ strikes the cylinder near its center (neglect rotational effects) with an initial speed of 90 km/hr. The initial velocity of $m_2$ is downward at a 45° angle in a vertical plane, which makes an acute angle $\phi$ relative to a plane tangent to $m_1$ and containing the contact point. Use an isotropic friction coefficient of 0.45 and a coefficient of restitution $e = 0.95$. Determine the final velocity components for $m_1 = 40$ kg, $m_2 = 1$ kg, and $\phi = 60°$.

*Solution:* From the illustration, the initial velocities in km/hr are seen to be

$$v_{1n} = 0, \qquad v_{2n} = -v_2 \cos 45° \sin \phi = -55.1$$

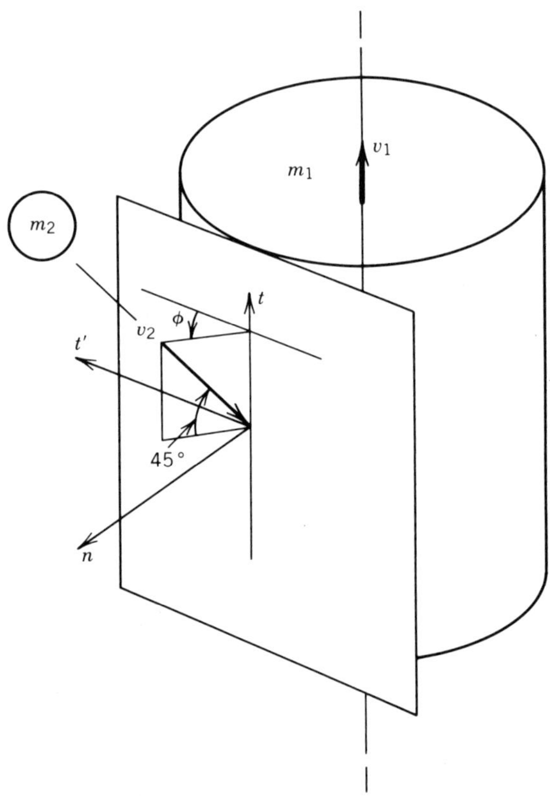


**Figure E4.1.** Particle–cylinder impact configuration.

$$v_{1t} = 300, \qquad v_{2t} = -v_2 \cos 45^\circ = -63.6$$
$$v_{1t'} = 0, \qquad v_{2t'} = -v_2 \cos 45^\circ \cos \phi = -35.4$$

Using these, the initial velocity ratios, $r$ and $r'$, can be calculated; these are $r = 6.60$ and $r' = 0.64$. For $e = 0.95$, the corresponding critical impulse ratios are $\mu_c = 3.38$ and $\mu_{c'} = 0.33$. With uniform friction properties, the critical impulse ratio is 3.40, since

$$\mu_c = \sqrt{\mu_c^2 + \mu_{c'}^2} = 3.40$$

Since $f < \mu_c$, it can be concluded that sliding continues throughout the contact duration. The final velocities are then obtained from Eqs. 4.22–4.27 with $\mu = f$. These are

$$V_{1n} = 2.62 \qquad V_{2n} = 49.7$$
$$V_{1t} = 299.0 \qquad V_{2t} = -30.3$$
$$V_{1t'} = -0.834 \qquad V_{2t'} = -2.07$$

all with units of km/hr. ■

## 4.2 PROBLEMS

**4.1** Two particles of mass $m_1 = 3$ kg and $m_2 = 4$ kg collide. The coefficient of restitution is 0.8 and the initial velocity components in units of m/s are

$$v_{1n} = -10, \qquad v_{2n} = 20$$
$$v_{1t} = -6, \qquad v_{2t} = 0$$
$$v_{1t'} = 0, \qquad v_{2t'} = 5$$

The impulse ratios are $\mu_t = 0.10$ and $\mu_{t'} = 0.05$. Determine the

1. final velocity components,
2. final relative tangential velocity,
3. system kinetic energy loss,
4. maximum system energy loss (for $e = 0.8$).

**4.2** A flat, rigid, and massive metal surface moves with velocity components $v_{2n} = 5$ m/s, $v_{2t} = 0$, and $v_{2t'} = 0$. The surface is finish machined with microgrooves parallel to the $t$ axis such that the friction coefficients are $f = 0.1$ and $f' = 0.3$. A particle of mass $m_1 = 0.01$ kg hits the surface with velocity components $v_{1n} = -20$ m/s, $v_{1t} = -50$ m/s, and $v_{1t'} = 10$ m/s. Determine the final velocity of the particle if $e = 0.70$. Calculate the tangential impulse components.

***Answers:*** $P_t = 0.043$ N-s, $P_{t'} = -0.100$ N-s.

**4.3** Find an initial velocity $v_1$ of the cylinder of Example 4.1 such that sliding ends at the instant of separation.

***Answers:*** $-97.5 \leq v_1 \leq 29.7$ km/hr

**4.4** Show that Eqs. 4.29 and 4.30 correspond to $V_{1t} = V_{2t}$ and $V_{1t'} = V_{2t'}$.

## CHAPTER 5

# PLANAR RIGID BODY COLLISIONS

Up to this point, the system and solution equations of collision dynamics for point masses have been presented. Rotational velocities and rotational inertias have been neglected; they are included in this chapter. The term rigid body as used here denotes that rotational properties are included, not that the bodies' materials are inflexible. On the other hand, the body must retain its shape during the collision with an exception discussed later. As in the point-mass case, the coefficients $e$ and $\mu$ retain their dual role in controlling energy loss and impact kinematics. However, their influence must be defined with respect to the velocities and impulses at the region of contact. The kinematic coefficient of restitution now is defined as the ratio of the final to initial relative velocity at the contact region, normal to a common tangential line. It is strongly influenced by the local deformation. The process of generation of the common tangential forces must also be referred to the contact region. Most prior treatments of this subject assume that contact between the two colliding objects occurs at a point. Although this assumption is often legitimate, it is not valid for a general treatment. Allowing contact forces and impulses to be distributed requires the introduction of a moment and moment impulse at the contact region. If a sudden, permanent shape change takes place of one or both bodies, the before-and-after nature of the impact problem allows the shape change to be taken into account through dimensional and inertial property changes. This and the other topics mentioned previously are treated in this chapter.

The basic assumptions are the same as before, such as short contact durations and instantaneous velocity changes. Negligible rotational displacements during the collision must now be added as an assumption. This is more likely to be violated for noncentral collisions with large moment arms and should be kept in mind when applying this theory.

The general approach followed in the development of equations will be the same as for point masses. System equations are derived based on principles of impulse and momentum and on other concepts such as energy loss and kinematics through the use of coefficients. The impulse variables are eliminated, and the system equations are then solved to produce the solution equations that relate final to initial velocities through the system parameters. Once the solution equations are found, an expression for the energy loss can be determined and the unknown impulses can be computed. The influence of the coefficients on the energy loss must be examined in order to determine the appropriate value of $\mu$ to be used in the solution equations. As for point-mass collisions, examination of the kinetic energy loss leads to a limiting value of the impulse ratio $\mu$ called $\mu$-critical. Examination of the tangential force system and its relationship to $\mu_c$ is used to choose the appropriate value of $\mu$.

A quandary arises in the two-body planar collision problem concerning the number and source of system equations. Each of the two rigid bodies has three final velocity components for a total of six unknowns: $V_{1n}$, $V_{1t}$, $\Omega_1$, $V_{2n}$, $V_{2t}$, and $\Omega_2$. The $\Omega$'s are the final angular velocities. Consequently, six system equations are required. Four transfer directly from the point-mass problem; these are

1. conservation of linear momentum in the direction of the normal coordinate;
2. conservation of linear momentum in the direction of the tangential coordinate;
3. the concept of restitution;
4. the tangential interaction of the bodies.

Various accounts[1,2] of sources of equations have been proposed. Typically, conservation of angular momentum about the point of contact of *each* of the two bodies is used as a source of two equations in addition to those obtained from the four items listed previously. However, this implies an assumption that is not always valid and that is also unnecessary. This is the assumption that contact occurs at a "point." Recognition that the interaction of one body with another during an impact occurs over a surface and that forces are surface forces, not point forces, requires introduction of a moment impulse. Figure 5.1 shows corresponding free body diagrams of two arbitrarily shaped bodies during a collision. Other pertinent quantities such as coordinate systems and dimensions are also illustrated.

Points $C_1$ and $C_2$ in Fig. 5.1 represent points in each body that are considered to be common during the collision configuration and are considered as a single point, $C$. In the presence of significant surface deformation, the location of points $C_1$ and $C_2$ are not always known. Although unknown, they are well defined, since they represent the point of application of the resultant impulse of the forces of interaction. Since interaction forces are dynamic surface forces, in general they will have a nonzero moment about $C$. Furthermore, the impulse of this moment

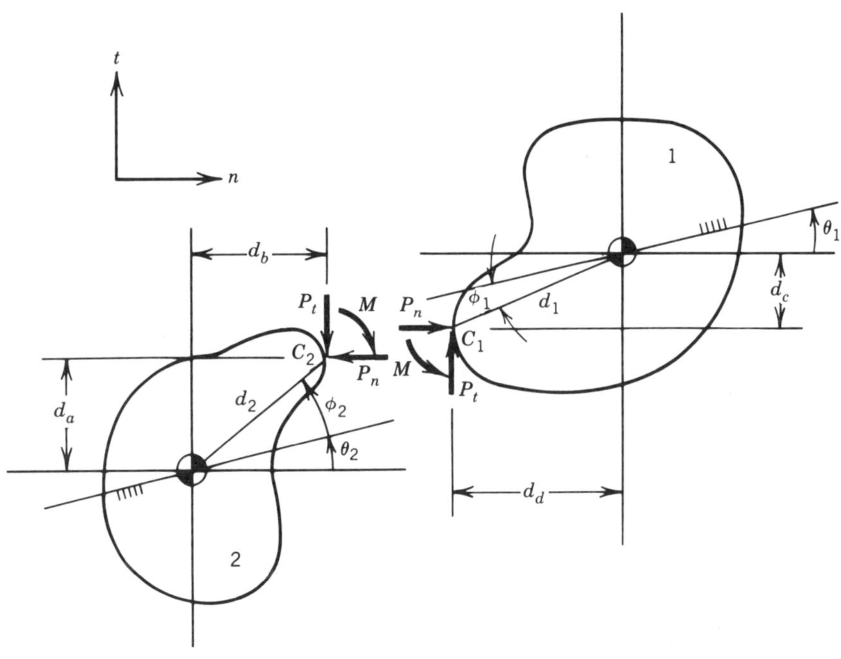


**Figure 5.1.** Free body diagram, planar rigid body impact.

may or may not be zero. Generality then requires that a moment impulse be permitted and is designated by $M$. This quantity is an unknown just as $P_n$ and $P_t$ are unknowns.

The existence of a moment impulse on each free body does not permit the use of conservation of angular momentum of each body individually about point $C$ as discussed earlier. In the absence of external forces and moments, angular momentum of the system or combination of masses still is conserved, however, and provides a fifth system equation. The absence of a sixth condition must be overcome. The approach used in the next section is to define a moment coefficient, $e_m$, with properties analogous to the classical coefficient of (normal) restitution $e$. The moment coefficient possesses properties that simultaneously control the rotational kinematics and influence the energy loss of the collision.

Linear momentum and angular momentum are conserved for the combined system of bodies in the absence of external forces and moments. For the special case of a collision of a body with a massive surface (or barrier), momentum is no longer conserved. So for a single rigid body colliding with a barrier, only the coefficient equations are used. This problem is covered in the next chapter.

As before, the initial velocities are considered as known and the final velocities are to be computed. Three final velocity components exist for each rigid body for motion in a plane for a total of six unknowns. After the system equations are found, momentum variables will be introduced and the equations will be placed into matrix form.

One of the assumptions made when using principles of impulse and momentum for modeling collisions is that the velocity changes of $m_1$ and $m_2$ are finite, whereas the displacements are small (infinitesimal). This assumption must now be extended to the angular positions of the rigid bodies (see Chapter 1). Other effects must be considered for rigid body problems. Rigid body properties such as dimensions $d_1$, $\phi_1$, $d_2$, and $\phi_2$ and moments of inertia $I_1$ and $I_2$ enter into the problem. Some of these dimensions may change significantly and permanently during the collision as a result of inelastic deformations. Recall that changes over a time interval from $\tau_1$ to $\tau_2$ are used in deriving equations for the system. These changes have involved momentum, velocity, and energy for mass particles. Equation 1.5 does not rule out changes in bodily configuration. As long as the changes are due to physical effects internal to the system only, external effects may continue to be neglected. Consequently, the dimensions and moments of inertia may have different final and initial values. Collisions accompanied by changes in bodily dimensions are covered later in this chapter.

## 5.1 SYSTEM EQUATIONS

In the following equations, it is assumed that all distances and angles in Fig. 5.1 are the same throughout the duration of contact. Conservation of linear momentum of the system in the direction of the normal coordinate $n$ can be written as (see Fig. 5.1)

$$m_1 V_{1n} + m_2 V_{2n} = m_1 v_{1n} + m_2 v_{2n} \tag{5.1}$$

Likewise, conservation of linear momentum in the tangential direction is

$$m_1 V_{1t} + m_2 V_{2t} = m_1 v_{1t} + m_2 v_{2t} \tag{5.2}$$

Note that all velocities in these and the following equations, unless otherwise indicated, are mass-center velocities. The final and initial relative normal velocity components are related by the coefficient of restitution. For rigid bodies, the relationship is defined for velocity components at the point of contact $C$. This can be written as

$$V_{1n} + d_c\Omega_1 - V_{2n} + d_a\Omega_2 = -e(v_{1n} + d_c\omega_1 - v_{2n} + d_a\omega_2) \tag{5.3}$$

where

$$d_a = d_2 \sin(\theta_2 + \phi_2) \quad \text{and} \quad d_c = d_1 \sin(\theta_1 + \phi_1) \tag{5.4}$$

It is possible here to use the kinetic definition of the coefficient of restitution, Eq. 1.14. For point-mass collisions, it was shown that these definitions yield identical results. In the case of rigid body collisions, this is no longer true (see problems 6.5 and 6.6). The kinematic definition, Eq. 5.3, is used in the remainder of this book. Investigation of the equivalence of the alternative definitions is relegated to future study.

The angles $\theta_1$ and $\theta_2$ are measured to fixed body axes. The angles $\phi_1$ and $\phi_2$ are relative to these body fixed axes and a line to the point $C$.

Similarly,

$$d_b = d_2 \cos(\theta_2 + \phi_2) \qquad \text{and} \qquad d_d = d_1 \cos(\theta_1 + \phi_1) \tag{5.5}$$

Also for subsequent use, note that the linear impulse components can be written as

$$P_n = \left[m_1(V_{1n} - v_{1n}) - m_2(V_{2n} - v_{2n})\right]/2 \tag{5.6}$$

and

$$P_t = \left[m_1(V_{1t} - v_{1t}) - m_2(V_{2t} - v_{2t})\right]/2 \tag{5.7}$$

The particular form for these equations is chosen to introduce a desirable symmetry into the equations. As in the point-mass problem, an impulse ratio coefficient can be defined as by Eq. 2.8. Using Eqs. 5.6 and 5.7, this gives

$$\begin{aligned} -\mu m_1 V_{1n} + m_1 V_{1t} + \mu m_2 V_{2n} - m_2 V_{2t} \\ = -\mu m_1 v_{1n} + m_1 v_{1t} + \mu m_2 v_{2n} - m_2 v_{2t} \end{aligned} \tag{5.8.1}$$

Keep in mind that Eq. 5.8.1 implies sliding throughout the impact. If sliding ends at or before separation, then

$$(V_{2t} + d_b\Omega_2) - (V_{1t} - d_d\Omega_1) = 0 \tag{5.8.2}$$

As in a point-mass collision problem, it will be noted that Eq. 5.8.2 provides a solution that is a special case of Eq. 5.8.1. Consequently, Eq. 5.8.1 is sufficient for the system equations.

A fifth equation is obtained from conservation of angular momentum of the system. The moment of momentum of the system about a point midway between the mass centers is set to zero. This gives

$$\begin{aligned} -d_{13}m_1 V_{1n} + d_{24}m_1 V_{1t} + d_{13}m_2 V_{2n} - d_{24}m_2 V_{2t} + I_1\Omega_1 + I_2\Omega_2 \\ = -d_{13}m_1 v_{1n} + d_{24}m_1 v_{1t} + d_{13}m_2 v_{2n} - d_{24}m_2 v_{2t} + I_1\omega_1 + I_2\omega_2 \end{aligned} \tag{5.9}$$

where

$$d_{13} = (d_a + d_c)/2 \qquad \text{and} \qquad d_{24} = (d_b + d_d)/2 \tag{5.10}$$

At this time, a sixth equation is necessary. With the potential existence of a moment in the contact region, it is possible that rotational velocities could be constrained as may be required in certain applications. Furthermore, a moment could lead to an energy loss if it has an inelastic behavior (no restoring moment, for example). For these reasons, it is reasonable to introduce a coefficient $e_m$, called the moment coefficient, to relate initial and final angular velocities. The equation could take a form analogous to Eq. 2.6, that is, $\Omega_2 - \Omega_1 = -e_m(\omega_2 - \omega_1)$, but this is too restrictive. In particular, this form of an equation requires that a moment be present for all collisions, an undesirable condition. To determine an appropriate form refer to Eq. 2.23, which shows that for $e = 0$, $V_{2n} = V_{1n}$ and $e = 1$ corresponds to a perfectly elastic impact. Note, however, that for the unrealistic value of $e = -1$, $P_n = 0$, which may be interpreted as no normal impulse. The same form as Eq. 2.23 is used to define the moment coefficient $e_m$; that is,

$$e_m M = -(1 + e_m)\bar{I}(\Omega_2 - \Omega_1) \tag{5.11}$$

where $\bar{I} = I_1 I_2/(I_1 + I_2)$. Note that for $e_m = 0$, $\Omega_2 = \Omega_1$, which corresponds to a perfectly inelastic angular impact, independent of the normal coefficient $e$. When $e_m = -1$, $M = 0$ and corresponds to a rigid body collision where the moment impulse is zero or can be neglected. That $e_m = 1$ corresponds to a perfectly elastic angular impact will be seen from the subsequent equations. Summing moments about the mass centers of each rigid body gives

$$I_1(\Omega_1 - \omega_1) = P_n d_c - P_t d_d + M \tag{5.12}$$

and

$$I_2(\Omega_2 - \omega_2) = P_n d_a - P_t d_b - M \tag{5.13}$$

From these,

$$M/I_1 = (\Omega_1 - \omega_1) - P_n d_c/I_1 + P_t d_d/I_1 \tag{5.14}$$

and

$$M/I_2 = -(\Omega_2 - \omega_2) + P_n d_a/I_2 - P_t d_b/I_2 \tag{5.15}$$

Equations 5.14 and 5.15 are used to eliminate $M$ from Eq. 5.11. Using Eqs. 5.6 and 5.7 to eliminate the impulses provides the sixth system equation. That is,

$$\begin{aligned} -e_m\gamma m_1 V_{1n} + e_m\delta m_1 V_{1t} + e_m\gamma m_2 V_{2n} - e_m\delta m_2 V_{2t} - \Omega_1 + \Omega_2 \\ = -e_m\gamma m_1 v_{1n} + e_m\delta m_1 v_{1t} + e_m\gamma m_2 v_{2n} \\ - e_m\delta m_2 v_{2t} + e_m\omega_1 - e_m\omega_2 \end{aligned} \tag{5.16}$$

where

$$\gamma = \left(\frac{d_c}{I_1} - \frac{d_a}{I_2}\right)\Big/2 \quad \text{and} \quad \delta = \left(\frac{d_d}{I_1} - \frac{d_b}{I_2}\right)\Big/2 \tag{5.17}$$

Equations 5.1–5.3, 5.8.1, 5.9, and 5.16 are the system equations. They form six linear algebraic equations whose solution furnishes the six unknown final velocity components of a rigid body collision. These can be written in a convenient matrix form using momentum variables. Let $U_{in} = m_i(V_{in} - v_{in})$, $U_{it} = m_i(V_{it} - v_{it})$, and $W_i = I_i(\Omega_i - \omega_i)$, $i = 1, 2$. The rigid body system equations become

$$\begin{bmatrix} 1 & 0 & 1 & 0 & 0 & 0 \\ 0 & 1 & 0 & 1 & 0 & 0 \\ 1/m_1 & 0 & -1/m_2 & 0 & d_c/I_1 & d_a/I_2 \\ -\mu & 1 & \mu & -1 & 0 & 0 \\ -d_{13} & d_{24} & d_{13} & -d_{24} & 1 & 1 \\ -e_m\gamma & e_m\delta & e_m\gamma & -e_m\delta & -1/I_1 & 1/I_2 \end{bmatrix} \begin{Bmatrix} U_{1n} \\ U_{1t} \\ U_{2n} \\ U_{2t} \\ W_1 \\ W_2 \end{Bmatrix}$$

$$= \begin{Bmatrix} B_1 \\ B_2 \\ B_3 \\ B_4 \\ B_5 \\ B_6 \end{Bmatrix} = \begin{Bmatrix} 0 \\ 0 \\ -(1 + e)(v_{1n} - v_{2n} + d_c\omega_1 + d_a\omega_2) \\ 0 \\ 0 \\ (1 + e_m)(\omega_1 - \omega_2) \end{Bmatrix} \tag{5.18}$$

where $B_1$ through $B_6$ are defined by the last column vector of Eq. 5.18. Note that these system equations reduce to the point-mass equations of Chapter 3 when all of the rigid body dimensions are zero and the rotational velocities vanish.

## 5.2 SOLUTION EQUATIONS

Equation 5.18 is not difficult to solve. Let each of the individual component equations (rows) of the matrix set be referred to as (a), (b), . . . , (f), respectively. Multiplying (c) by $m_2$ and adding to (a) gives

$$U_{1n} = \overline{m}B_3 - (\overline{m}d_c/I_1)W_1 - (\overline{m}d_a/I_2)W_2 \qquad (5.19)$$

Multiplying (d) by $d_{24}$, subtracting from (e), and using Eq. 5.19 gives

$$\begin{aligned}&[1 + 2(d_{13} - \mu d_{24})(\overline{m}d_c/I_1)]W_1 \\ &\quad + [1 + 2(d_{13} - \mu d_{24})(\overline{m}d_a/I_2)]W_2 \\ &= 2(d_{13} - \mu d_{24})\overline{m}B_3 \qquad (5.20)\end{aligned}$$

Multiplying (d) by $e_m\delta$, subtracting from (f), and using Eq. 5.19 gives

$$\begin{aligned}&[-1 + 2e_m(\gamma - \mu\delta)\overline{m}d_c]W_1/I_1 + [1 + 2e_m(\gamma - \mu\delta)\overline{m}d_a]W_2/I_2 \\ &\quad = B_6 + 2e_m(\gamma - \mu\delta)\overline{m}B_3 \qquad (5.21)\end{aligned}$$

Equations 5.20 and 5.21 are two independent simultaneous equations for $W_1$ and $W_2$. Their solution can be used with Eq. 5.19 to provide $U_{1n}$ and with (a) to provide $U_{2n}$. Written in the form of final velocities, these are

$$\begin{aligned}V_{1n} &= v_{1n} + \overline{m}(1 + e)(v_{2n} - v_{1n})/m_1 - \overline{m}[d_c(\Omega_1 + e\omega_1) \\ &\quad + d_a(\Omega_2 + e\omega_2)]/m_1 \qquad (5.22)\end{aligned}$$

and

$$\begin{aligned}V_{2n} &= v_{2n} - \overline{m}(1 + e)(v_{2n} - v_{1n})/m_2 + \overline{m}[d_c(\Omega_1 + e\omega_1) \\ &\quad + d_a(\Omega_2 + e\omega_2)]/m_2 \qquad (5.23)\end{aligned}$$

The fourth of the matrix equations, (d), shows that $U_{1t} = \mu U_{1n}$ and $U_{2t} = \mu U_{2n}$. Expressed in final velocities, these are

$$V_{1t} = v_{1t} + \mu\overline{m}(1 + e)(v_{2n} - v_{1n})/m_1 - \mu\overline{m}[d_c(\Omega_1 + e\omega_1) + d_a(\Omega_2 + e\omega_2)]/m_1 \tag{5.24}$$

$$V_{2t} = v_{2t} - \mu\overline{m}(1 + e)(v_{2n} - v_{1n})/m_2 + \mu\overline{m}[d_c(\Omega_1 + e\omega_1) + d_a(\Omega_2 + e\omega_2)]/m_2 \tag{5.25}$$

Equations 5.22–5.25 contain the final angular velocities as obtained from solution of Eqs. 5.20 and 5.21. Among other variables, $\Omega_1$ and $\Omega_2$ will be functions of $e_m$, $I_1$, and $I_2$, which do not appear explicitly in Eqs. 5.22–5.25.

The solution of Eqs. 5.20, 5.21, and 5.22–5.25 are the solution equations for the planar, rigid body collision problem. They contain three coefficients: $e$, $\mu$, and $e_m$. As before, the coefficient of restitution $e$ plays a dual role as a kinematic coefficient governing velocity ratios and an energy-loss coefficient representing inelastic deformation. Classically, $0 \leq e \leq 1$; however, negative values of $e$ may provide realistic results where penetration rather than rebound occurs. No general proof has been presented that the upper bound of 1 on $e$ follows from energy principles. Tangential effects are again modeled using the coefficient $\mu$. As seen in the point-mass solution, the value of $\mu$ can be either positive or negative. Its appropriate sign will be determined later, based on concepts of energy loss and kinematics.

The value of $\mu$ used in the solution equations influences the energy lost during the collision. As in the point-mass problem, the choice of $\mu$ is not completely arbitrary; improperly chosen values can lead to unrealistic solutions and violation of energy-conservation principles. The coefficient $\mu$ also plays a multiple role. It can be used as a friction coefficient when appropriate, which also influences the energy loss, and it also can be used to control the relative tangential velocities. A unique value of $\mu$ exists that corresponds to the condition of zero relative final tangential velocity at point $C$; call this $\mu_0$. It can be found by using Eq. 5.8.2 in place of Eq. 5.8.1 in the system equations. The value of $\mu$ from this solution is $\mu_0$. That is,

$$\mu_0 = P_t/P_n\big|_{\text{Eq. 5.8.1}} \tag{5.26}$$

is the value of the impulse ratio corresponding to the kinematic condition that the final relative tangential velocity at point $C$ is zero.

The newly defined moment coefficient also plays a multifaceted role. Recall that for $e_m = -1$ the moment impulse $M = 0$. For $0 \leq e_m \leq 1$, the equations model an elastic, angular impact with a reversal of sign of the relative angular velocities. For negative values of $e_m$, that is, $-1 < e_m < 0$, an "angular penetration" can be simulated. This feature must be used with care since values of $e_m$ can exist for which the determinant of Eq. 5.20 and 5.21 becomes zero.

The solution equations contain the three coefficients: $e$, $e_m$, and $\mu$. Before values are chosen for these coefficients for applications it is necessary to consider how the kinetic energy loss of the collision is affected by their values and how their values are related to the physical process of generation of forces and moments during the contact duration. This must be done to ensure that energy-conservation laws are not violated and that the tangential impulse (through the sign and magnitude of $\mu$) and moment impulse (through the sign and magnitude of $e_m$) are consistent with the initial and final collision kinematics.

Examples and more discussion of these coefficients follow.

## ■ *Example 5.1*

PROBLEM: Matched pairs of colliding parts are shown. The physical properties of the parts are:

$$m_1 = 8 \times 10^{-3} \text{ lb-s}^2/\text{in.}, \qquad I_1 = 40 \times 10^{-3} \text{ in.-lb-s}^2$$

$$m_2 = 4 \times 10^{-3} \text{ lb-s}^2/\text{in.}, \qquad I_2 = 80 \times 10^{-3} \text{ in.-lb-s}^2$$

Determine the final angular velocities of $m_1$ and $m_2$ for each pair such that the pegs remain engaged in their respective slots following impact and for the following initial conditions:

$$v_{1n} = 5 \text{ in./s}, \qquad v_{2n} = 2 \text{ in./s}$$

$$v_{1t} = 0, \qquad v_{2t} = 2 \text{ in./s}$$

$$\omega_1 = 0, \qquad \omega_2 = -3 \text{ rad/s}$$

$$d_1 = 3 \text{ in.}, \qquad d_2 = 3 \text{ in.}$$

$$\phi_1 = 60^\circ, \qquad \phi_2 = 60^\circ$$

$$\theta_1 = 0, \qquad \theta_2 = 0$$

*Solution:* Since the pegs remain engaged in the slots, the collisions of both pairs are perfectly inelastic and $e = 0$ is used. Since the system equations are rather long, a computer solution was used to obtain the following results.

Case 1, Round Peg: For this configuration, no impulse moment at the contact area is permitted, so $e_m = -1$. Under this condition the corre-

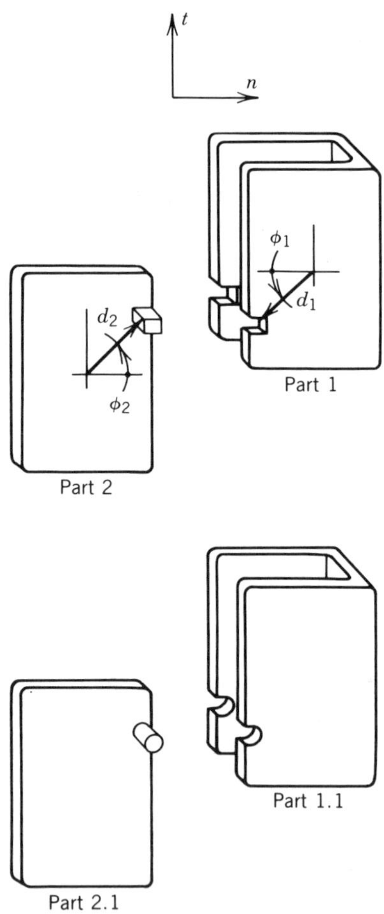


**Figure E5.1.** Colliding machine parts. Mating parts 1 and 2 give zero final relative rotation; mating parts 1.1 and 2.1 have free final relative rotation.

sponding value of $\mu_0$ (see Eq. 5.8.2) is $-0.427$. For these values of coefficients, the solution equations give

$$\Omega_1 = 0.386 \text{ rad/s}, \qquad \Omega_2 = -2.23 \text{ rad/s}$$

Case 2, Square Peg: Here, the collision has a zero final relative angular velocity that corresponds to a perfectly inelastic angular impact with $e_m = 0$. The corresponding value of $\mu_0$ is $-0.285$. The solution equations now give

$$\Omega_1 = \Omega_2 = -0.754 \text{ rad/s}$$

■

## 5.3 SOLUTION EQUATIONS: ZERO MOMENT IMPULSE

An important special case occurs when the moment impulse $M = 0$, which corresponds to $e_m = -1$. This is the solution of the planar impact problem where the contact surface forces can be represented as point forces at a known point. In this case the system equations are simpler and an analytical solution can be found using the approach of Section 5.2. The solution equations are as follows:

$$V_{1n} = v_{1n} + \overline{m}(1 + e)[(v_{2n} - d_a\omega_2) - (v_{1n} + d_c\omega_1)]q/m_1 \tag{5.27}$$

$$V_{2n} = v_{2n} - \overline{m}(1 + e)[(v_{2n} - d_a\omega_2) - (v_{1n} + d_c\omega_1)]q/m_2 \tag{5.28}$$

$$V_{1t} = v_{1t} + \mu\overline{m}(1 + e)[(v_{2n} - d_a\omega_2) - (v_{1n} + d_c\omega_1)]q/m_1 \tag{5.29}$$

$$V_{2t} = v_{2t} - \mu\overline{m}(1 + e)[(v_{2n} - d_a\omega_2) - (v_{1n} + d_c\omega_1)]q/m_2 \tag{5.30}$$

$$\Omega_1 = \omega_1 + \overline{m}(1 + e)[(v_{2n} - d_a\omega_2) - (v_{1n} + d_c\omega_1)] \cdot (d_c - \mu d_d)d_e q/I_1 \tag{5.31}$$

$$\Omega_2 = \omega_2 + \overline{m}(1 + e)[(v_{2n} - d_a\omega_2) - (v_{1n} + d_c\omega_1)] \cdot (d_a - \mu d_b)d_f q/I_2 \tag{5.32}$$

where

$$q = \left[1 + \frac{\overline{m}d_a^2}{I_2} + \frac{\overline{m}d_c^2}{I_1} - \mu\left(\frac{\overline{m}d_c d_d}{I_1} + \frac{\overline{m}d_a d_b}{I_2}\right)\right]^{-1} \tag{5.33}$$

Although these equations are for $e_m = -1$, they provide a solution to an impact problem not completely solved in the past.

## 5.4 ENERGY LOSS: CRITICAL IMPULSE RATIO

As in the collision problem of two point masses, examination of the energy loss and the corresponding role of the coefficients in controlling energy loss is necessary. Three coefficients now exist: $e$, $e_m$, and $\mu$. Unfortunately, a single, concise expression for $T_L$ in terms of all three coefficients has not yet been derived. Analytical solutions can provide more efficient insight and so the special case of zero moment impulse is examined, since a corresponding expression for $T_L$ is obtainable. That is, the following applies for $M = 0$ and $e_m = -1$.

The energy loss $T_L$ can be found by its definition, which is

$$\begin{aligned} T_L = {} & \tfrac{1}{2}m_1(v_{1n}^2 + v_{1t}^2) + \tfrac{1}{2}m_2(v_{2n}^2 + v_{2t}^2) + \tfrac{1}{2}I_1\omega_1^2 + \tfrac{1}{2}I_2\omega_2^2 \\ & - \tfrac{1}{2}m_1(V_{1n}^2 + V_{1t}^2) - \tfrac{1}{2}m_2(V_{2n}^2 + V_{2t}^2) - \tfrac{1}{2}I_1\Omega_1^2 - \tfrac{1}{2}I_2\Omega_2^2 \end{aligned} \tag{5.34}$$

Substitution for the final velocities gives

$$\begin{aligned} T_L = {} & \tfrac{1}{2}\overline{m}(1 + e)qv_{rn}^2\Bigg[2 + 2\mu r - (1 + e) \\ & \cdot q\left(1 + \mu^2 + \frac{\overline{m}d_e^2}{I_1} + \frac{\overline{m}d_f^2}{I_2}\right)\Bigg] \end{aligned} \tag{5.35}$$

where

$$v_{rn} = (v_{2n} - d_a\omega_2) - (v_{1n} + d_c\omega_1) \tag{5.36}$$

$$d_e = d_c - \mu d_d \qquad \text{and} \qquad d_f = d_a - \mu d_b \tag{5.37}$$

$$r = \frac{(v_{2t} + d_b\omega_2) - (v_{1t} - d_d\omega_1)}{(v_{2n} - d_a\omega_2) - (v_{1n} + d_c\omega_1)} \tag{5.38}$$

and $q$ is given by Eq. 5.33.

The coefficients $e$ and $\mu$ are parameters that represent the physical processes in which energy is lost during a collision. Because these coefficients are also used as kinematic constraints, care must be taken to ensure that solutions are realistic. A kinematic constraint (for example, a value of $\mu$ to give a zero final tangential velocity) may be able to produce a set of final velocities that a process such as friction cannot physically achieve. Friction cannot do net positive work on a system of rigid bodies, yet the impact equations are capable of producing such solutions.

It is not difficult to develop an understanding of how $T_L$ is affected by changes in $e$ and $\mu$ for point-mass collisions. Establishing limiting or critical values of $\mu$ is somewhat more involved for rigid body collisions. As in the point-mass collision, $T_L(\mu)$ possesses a maximum. The impulse ratio for this maximum is denoted by $\mu = \mu_m$. Another special value of $\mu$ frequently needed is that which corresponds to zero final relative tangential velocity at the contact area. This was discussed earlier and is designated as $\mu_0$. In the point-mass-collision problem, zero final tangential motion and maximum energy loss occur for the same impulse ratio. This was defined as the critical impulse ratio. For rigid body collisions $\mu_0$ and $\mu_m$ are generally different, since noncentral collision geometry plays an important role. Finally, $\mu_{max}$ will denote the value of $\mu_m$ for the special case at $e = 0$, that is, $\mu_{max} = \mu_m(e = 0)$.

## ■ *Example 5.2*

PROBLEM: A slender rod with mass $m_1 = 1000$ is stuck on its tip by a sphere with mass $m_2 = 500$; they are oriented as shown. The initial velocities are

$$v_{1n} = -1, \qquad v_{2n} = 1$$

$$v_{1t} = -2, \qquad v_{2t} = 3$$

$$\omega_1 = 0, \qquad \omega_2 = 0$$

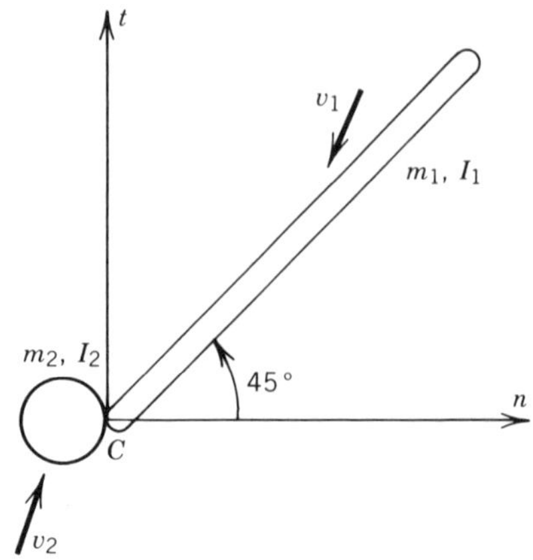


**Figure E5.2.1.** Impact of a sphere and rod end.

The physical characteristics of the system (see Fig. E5.1) are given by

$$\phi_1 = 0, \qquad \phi_2 = 0, \qquad I_1 = 83.33, \qquad I_2 = 0.02$$
$$\theta_1 = 45°, \qquad \theta_2 = 0, \qquad d_1 = 0.5, \qquad d_2 = 0.01$$

Treating these data as a consistent set of units, plot the energy loss (expressed as a percentage of initial system energy) as a function of impulse ratio for $e = 0, \frac{1}{2}$, and 1. The tangential force process is dry friction.

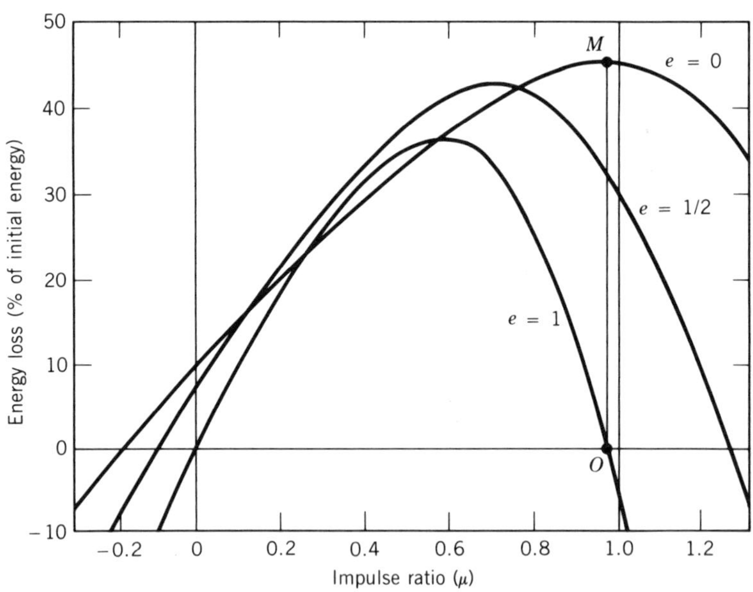


**Figure E5.2.2.** Energy loss.

*Solution:* Using the numerical data supplied, the energy loss, Eq. 5.35, becomes

$$T_L = \frac{1000(1 + e)}{2.25 - 0.75\mu}\left[2 + 5\mu - (1 + e)\frac{3.167\mu^2 - \mu + 1.5}{1.5 - 0.5\mu}\right]$$

This equation is shown plotted for the three values of $e$. Some pertinent comments can be made about the energy loss. First, the overall maximum loss is approximately 45% and occurs for $\mu = 0.93$; see point $M$.

Note that for $e = 1$ and $\mu < 0$ or $\mu > 0.93$ (see point $O$), the energy loss is negative. A negative value of $\mu$ can be used in the solution equations, but this requires the tangential impulse to have a direction opposite of that expected from frictional forces. Consequently, for $e = 1$, $\mu < 0$ requires the tangential impulse to do positive work on the system and increases its energy (negative energy loss). The solution corresponding to point $O$ is where the final tangential velocity at the contact point between the sphere and rod is zero. That is, sliding ends at separation and $\mu = \mu_0$. For $\mu > 0.93$, the tangential impulse is large enough to reverse sliding, which requires the addition of energy to the system already removed by sliding friction. This clearly is not possible from frictional forces. In the case of point-mass collisions, $\mu_c = \mu_m$, but here it is seen that $\mu_c = \mu_0 > \mu_m$. Note, finally that the curves intersect. This implies a greater energy loss for higher values of $e$ for some values of $\mu$, a behavior that is not obvious. ■

In order to establish a critical value of $\mu$ for rigid body collisions, some specific values are needed. Maximization of $T_L$ with respect to $\mu$ provides $\mu_m$, where

$$\mu_m(e) = \frac{\dfrac{r}{1 + e}A + \dfrac{1}{1 + e}B}{1 + C + \dfrac{r}{1 + e}B - \dfrac{e}{1 + e}\dfrac{B^2}{A}} \tag{5.39}$$

and

$$A = 1 + \frac{\overline{m}d_c^2}{I_1} + \frac{\overline{m}d_a^2}{I_2}, \quad B = \frac{\overline{m}d_c d_d}{I_1} + \frac{\overline{m}d_a d_b}{I_2}, \quad C = \frac{\overline{m}d_d^2}{I_1} + \frac{\overline{m}d_b^2}{I_2} \tag{5.40}$$

Note that by definition $\mu_{max} = \mu_m(e = 0)$ is

$$\mu_{max} = \frac{rA + B}{1 + C + rB} \tag{5.41}$$

If the system equations are solved using Eq. 5.8.2 instead of Eq. 5.8.1, then the solution corresponds to zero final relative tangential velocity. Using this solution, $P_t/P_n$ is set equal to $\mu_0$, and provides

$$\mu_0(e) = \frac{\frac{r}{1 + e} A + B}{1 + C + \frac{r}{1 + e} B} \tag{5.42}$$

Another special value of the impulse ratio is that when the energy loss $T_L$ is zero, $\mu = \mu_T$. Using Eq. 5.35, this is found to be

$$\mu_T(e) = \frac{rA + eB}{2rB + (1 + e)(1 + C)} \cdot \left\{ 1 \pm \left[ 1 + (1 - e)A \frac{2rB + (1 + e)(1 + C)}{(rA + eB)^2} \right]^{1/2} \right\} \tag{5.43}$$

Some observations can now be made about these special values of $\mu$. Note that

$$\mu_0(0) = \mu_T(1) = \mu_{max} = \frac{rA + B}{1 + C + rB} \tag{5.44}$$

For $B = 0$,

$$\mu_0(e) = \mu_m(e) = \frac{\frac{r}{1 + e} A}{1 + C} \tag{5.45}$$

Equation 5.44 indicates that the energy loss curve for perfectly elastic conditions, $e = 1$, crosses the horizontal axis at $\mu = \mu_T(1)$, the same value of $\mu$ for which the perfectly plastic energy loss curve reaches its maximum, which is also the global maximum, $T_{\mathrm{Lmax}}$. This is also the same value of $\mu$, $\mu_0(0)$, for zero final relative tangential velocity for $e = 0$. Equation 5.45, valid for $B = 0$, applies to central impacts of rigid bodies since $d_a = d_c = 0$. Consequently, for central collisions of rigid bodies (either direct or oblique) and for any value of $e$ the condition for zero final relative tangential velocity at the point of contact, $\mu = \mu_0$, is identical to the value of $\mu$ which maximizes each energy-loss curve. This is identical to what was found for point-mass impacts.

If the process of generation of the tangential force is passive (such as for frictional forces), the tangential impulse cannnot do positive work on the system. The local tangential forces can only oppose the relative velocity. Consequently, the impulse ratio can never exceed a value that corresponds to $|\mu_0|$. Moreover, a passive system can never gain energy due to an impact, so that $|\mu|$ can never exceed $|\mu_T|$. Of course a friction coefficient $f$, if appropriate, can be less than these limiting values. Under these conditions and for a given value of $e$, $\mu$ is limited by a critical value $\mu_c$, that is,

$$|\mu| \leq |\mu_c| = \min[f, |\mu_0|, |\mu_T|] \tag{5.46}$$

where

$$\mathrm{sgn}(\mu) = \mathrm{sgn}(\mu_{\max}) \tag{5.47}$$

Equations 5.46 and 5.47 are interpreted in the following manner. Under no circumstances can the value of $|\mu|$ used in the solution equations exceed $|\mu_T|$ without introducing energy into the system. If the tangential force is due to dry friction with a coefficient $f$, $|\mu|$ is limited by the smallest of $f$, $|\mu_0|$, and $|\mu_T|$. If $B = 0$, $|\mu|$ is bounded by $|\mu_0|$, which is also the case for point-mass collisions (where $B = C = 0$ and $A = 1$). When determining $\mu_c$ it should also be noted that for any amount of friction, the final relative tangential velocity should never be greater in magnitude than for the same, but frictionless, impact. That is, additional friction can reduce the magnitude of the final velocity (from the frictionless value), but never increase it.

Equation 5.47 provides the appropriate sign for $\mu$, which depends on initial conditions and the kinematics of the collision. It is important to note that cessation of sliding does not necessarily imply a maximum energy loss.

Because of the wide variety of physical systems, the energy loss must be examined in each case to see if the solutions are realistic and whether or not energy conservation principles are violated. More examples will be considered in Chapter 6 for barrier collisions.

## 5.5 COLLISIONS ACCOMPANIED BY SHAPE CHANGES

If, at the time of separation, the shape of either or both rigid bodies differs significantly from its shape at the beginning of contact, the previous system equations are not appropriate. This is because angular momentum depends on the body dimensions and moments of inertia as well as the angular velocities. Because of the before-and-after nature of the impact equations, these changes can be taken into account. Changes in mass are not considered, however. Shape changes still taking place dynamically at separation also are not considered, such as large amplitude flexible oscillations.

Again consider the rigid bodies depicted in Fig. 5.1. The quantities that can change are $d_a$, $d_b$, $d_c$, $d_d$ (or $d_1$, $d_2$, $\phi_1$, and $\phi_2$) and corresponding changes in $I_1$ and $I_2$.* The linear momentum equations, 5.1 and 5.2, remain unchanged. The concept of restitution is provided by Eq. 5.3; it now becomes

$$V_{1n} + D_c\Omega_1 - V_{2n} + D_a\Omega_2 = -e(v_{1n} + d_c\omega_1 - v_{2n} + d_a\omega_2) \tag{5.48}$$

where

$$D_a = D_2 \sin(\Theta_2 + \Phi_2) \quad \text{and} \quad D_c = D_1 \sin(\Theta_1 + \Phi_1) \tag{5.49}$$

and

$$D_b = D_2 \cos(\Theta_2 + \Phi_2) \quad \text{and} \quad D_d = D_1 \cos(\Theta_1 + \Phi_1) \tag{5.50}$$

Equation 5.8.1 results from the definition of $\mu$, and remains unchanged. The system angular momentum must remain constant in the absence of any external moments. Here, the angular momentum is ex-

*As for velocities, the final values of dimensions will be designated by capital symbols. An exception is moments of inertia; final values are given the symbol $J$ in place of $I$.

pressed again about the midpoint between the original mass-center positions. This gives a fifth system equation,

$$\begin{aligned}-D_{13}m_1V_{1n} &+ D_{24}m_1V_{1t} + D'_{13}m_2V_{2n}\\ &- D'_{24}m_2V_{2t} + J_1\Omega_1 + J_2\Omega_2\\ &= -d_{13}m_1v_{1n} + d_{24}m_1v_{1t} + d_{13}m_2v_{2n}\\ &- d_{24}m_2v_{2t} + I_1\omega_1 + I_2\omega_2\end{aligned} \tag{5.51}$$

where

$$\begin{aligned}D_{13} &= (d_a + d_c)\left(\frac{\overline{m}}{m_2} - \frac{1}{2}\right) + (D_a + D_c)\left(1 - \frac{\overline{m}}{m_2}\right)\\ D'_{13} &= (D_a + D_c) - D_{13}\end{aligned} \tag{5.52.1}$$

and

$$\begin{aligned}D_{24} &= (d_b + d_d)\left(\frac{\overline{m}}{m_2} - \frac{1}{2}\right) + (D_b + D_d)\left(1 - \frac{\overline{m}}{m_2}\right)\\ D'_{24} &= (D_b + D_d) - D_{24}\end{aligned} \tag{5.52.2}$$

The quantities $d_{13}$ and $d_{24}$ are defined in Eq. 5.10.

Equation 5.11 is retained as a definition of the moment coefficient $e_m$. To determine an expression for the moment impulse $M$ (in order to eliminate it), it is necessary to draw a distinction between the compression (or approach) phase of the impact and the expansion (or rebound) phase. Let $\overline{\tau}$ be the time during the contact duration, $\tau_2 - \tau_1$, at which the compression at point $C$ ends and rebound begins. The normal impulse during the compression phase, $\tau_1$ to $\overline{\tau}$, is $P_n/(1 + e)$. During the rebound phase, $\overline{\tau}$ to $\tau_2$, it is $eP_n/(1 + e)$. During compression the tangential impulse is written as $P_t/(1 + e)$ and as $eP_t/(1 + e)$ during rebound. *If sliding ends early in the collision duration, this partitioning of the tangential impulse may not be valid. Consequently, the following applies to central collisions or where sliding exists at separation or ends at or near separation.* (See Problems 6.5 and 6.6.)

An assumption now is made that shape changes of bodies 1 and 2 take place instantaneously at the end of the compression phase, just as $\tau =$

$\bar{\tau}$. Other possibilities exist and may be considered by the reader. As a consequence of these assumptions, the angular momentum equations can be found. The change in angular momentum of body 1 about its mass center is equal to the net moment of the impulses. For the compression or approach phase this is

$$J_1\bar{\omega}_1 - I_1\omega_1 = (1 - \lambda)M + d_cP_n/(1 + e) - d_dP_t/(1 + e) \tag{5.53}$$

where $\bar{\omega}_1$ is the angular velocity of $m_1$ at $\tau = \bar{\tau}$ and $\lambda$ is an arbitrary constant, such that $0 \leq \lambda \leq 1$. For the rebound phase

$$J_1\Omega_1 - J_1\bar{\omega}_1 = \lambda M + D_c eP_n/(1 + e) - D_d eP_t/(1 + e) \tag{5.54}$$

Adding this equation to Eq. 5.53 gives

$$J_1\Omega_1 - I_1\omega_1 = M + \frac{P_n}{1 + e}(d_c + eD_c) - \frac{P_t}{1 + e}(d_d + eD_d) \tag{5.55}$$

Similarly for body 2:

$$J_2\Omega_2 - I_2\omega_2 = -M + \frac{P_n}{1 + e}(d_a + eD_a) - \frac{P_t}{1 + e}(d_b + eD_b) \tag{5.56}$$

The quantities $M/I_1$ and $M/I_2$ can be obtained from Eqs. 5.55 and 5.56 and $M/\bar{I}$ eliminated from Eq. 5.11. This gives the sixth system equation,

$$\begin{aligned} &-e_m m_1 \Gamma V_{1n} + e_m m_1 \Delta V_{1t} + e_m m_2 \Gamma V_{2n} - e_m m_2 \Delta V_{2t} \\ &+ [e_m(J_1/I_1 - 1) - 1]\Omega_1 - [e_m(J_2/I_2 - 1) - 1]\Omega_2 \\ &\quad = -e_m m_1 \Gamma v_{1n} + e_m m_1 \Delta v_{1t} + e_m m_2 \Gamma v_{2n} \\ &\quad - e_m m_2 \Delta v_{2t} + e_m \omega_1 - e_m \omega_2 \end{aligned} \tag{5.57}$$

where

$$2\Gamma = \frac{d_c + eD_c}{(1 + e)I_1} - \frac{d_a + eD_a}{(1 + e)I_2}$$

and

$$2\Delta = \frac{d_d + eD_d}{(1 + e)I_1} - \frac{d_b + eD_b}{(1 + e)I_2} \tag{5.58}$$

Equations 5.1, 5.2, 5.48, 5.8.1, 5.51, and 5.57 are the system equations for a collision accompanied by a permanent change in shape of the rigid bodies. They are identical to the previous system equations for no significant changes in shape.

### ■ *Example 5.3*

PROBLEM: A hard spherical mass ($m_1$, $I_1$) collides with the center of a soft pliable rod ($m_2$, $I_2$). Owing to the collision, the rod deforms permanently making equal angles $\beta$ with its original orientation as shown. The masses stick together and move as a single rigid body following the collision. Determine the final velocities of the two rigid bodies if the rod is initially at rest and the sphere has initial velocities given by

$$v_{1n} = -v_0 \cos \phi, \quad v_{1t} = -v_0 \sin \phi, \quad \text{and} \quad \omega_1 = 0$$

also

$$m_1 = m = 3m_2/2, \quad L \sin \beta = 4R, \quad \text{and} \quad L = 10R$$

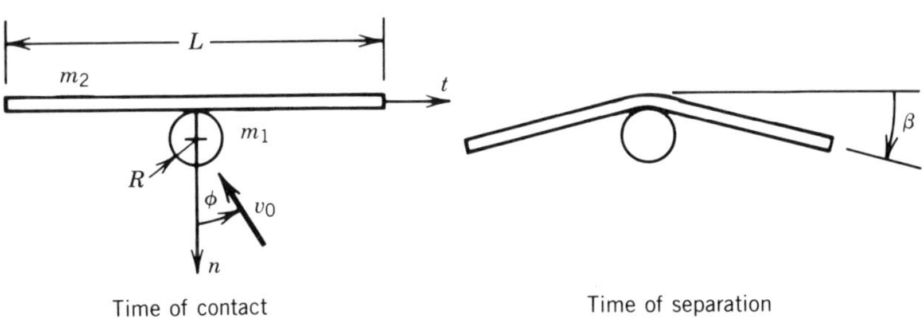


**Figure E5.3.** System configurations at time of contact and time of separation.

*Solution:* A summary of the system's constants are provided in the following table:

| Physical Constants | | Initial Values | Final Values |
|---|---|---|---|
| $m_1$ | $m_2$ | $m, 2m/3$ | $m, 2m/3$ |
| $I_1$ | $J_1$ | $mR^2/2$ | $mR^2/2$ |
| $I_2$ | $J_2$ | $mL^2/18$ | $mL^2(1 + 3\cos^2\beta)/72$ |
| $d_a$ | $D_a$ | 0 | 0 |
| $d_b$ | $D_b$ | 0 | $-L(\sin\beta)/4$ |
| $d_c$ | $D_c$ | 0 | 0 |
| $d_d$ | $D_d$ | $R$ | $R$ |
| $2\gamma$ | $2\Gamma$ | 0 | 0 |
| $2\delta$ | $2\Delta$ | $\frac{2}{mR}$ | $\frac{2}{mR} + \frac{9e\sin\beta}{2(1+e)mL}$ |
| $d_{13}$ | $D_{13}, D'_{13}$ | 0 | 0, 0 |
| $d_{24}$ | $D_{24}, D'_{24}$ | $\frac{R}{2}$ | $\frac{R}{2} - \frac{2L\sin\beta}{20}, \frac{R}{2} - \frac{3L\sin\beta}{20}$ |

Since the system behaves as a single rigid body at the end of the collision, the coefficients $e$ and $e_m$ are both zero. Also, for the same reason, Eq. 5.8.1 is replaced by Eq. 5.8.2. The system equations are

$$m_1 V_{1n} + m_2 V_{2n} = -m_1 v_0 \cos\phi \quad \text{(a)}$$

$$m_1 V_{1t} + m_2 V_{2t} = -m_1 v_0 \sin\phi \quad \text{(b)}$$

$$V_{1n} - V_{2n} = 0 \quad \text{(c)}$$

$$V_{1t} - R\Omega_1 - V_{2t} - D_b\Omega_2 = 0 \quad \text{(d)}$$

$$D_{24} m_1 V_{1t} - D'_{24} m_2 V_{2t} + J_1\Omega_1 + J_2\Omega_2 = -d_{24} m_1 v_0 \sin\phi \quad \text{(e)}$$

$$\Omega_1 - \Omega_2 = 0 \quad \text{(f)}$$

Equations a and c can be solved independently to give

$$V_{1n} = V_{2n} = -\tfrac{3}{5} v_0 \cos\phi \quad \text{(g)}$$

From Eq. f it is seen that $\Omega_1 = \Omega_2 = \Omega$. Equations b, d, and e can then be solved to furnish

$$V_{1t} = V_{2t} = -\tfrac{3}{5} v_0 \sin \phi \tag{h}$$

and

$$\Omega = -\tfrac{36}{485} v_0 \sin \phi \tag{i}$$

■

## 5.6 SUMMARY OF EQUATIONS, RIGID BODY COLLISION, ZERO MOMENT IMPULSE, $e_m = -1$

*Solution Equations*

$$V_{1n} = v_{1n} + \overline{m}(1+e)[(v_{2n} - d_a\omega_2) - (v_{1n} + d_c\omega_1)]q/m_1 \tag{5.27}$$

$$V_{2n} = v_{2n} - \overline{m}(1+e)[(v_{2n} - d_a\omega_2) - (v_{1n} + d_c\omega_1)]q/m_2 \tag{5.28}$$

$$V_{1t} = v_{1t} + \mu\overline{m}(1+e)[(v_{2n} - d_a\omega_2) - (v_{1n} + d_c\omega_1)]q/m_1 \tag{5.29}$$

$$V_{2t} = v_{2t} - \mu\overline{m}(1+e)[(v_{2n} - d_a\omega_2) - (v_{1n} + d_c\omega_1)]q/m_2 \tag{5.30}$$

$$\Omega_1 = \omega_1 + \overline{m}(1+e)[(v_{2n} - d_a\omega_2) - (v_{1n} + d_c\omega_1)]d_e q/I_1 \tag{5.31}$$

$$\Omega_2 = \omega_2 + \overline{m}(1+e)[(v_{2n} - d_a\omega_2) - (v_{1n} + d_c\omega_1)]d_f q/I_2 \tag{5.32}$$

$$q = \left[1 + \frac{\overline{m}d_a^2}{I_2} + \frac{\overline{m}d_c^2}{I_1} - \mu\left(\frac{\overline{m}d_c d_d}{I_1} + \frac{\overline{m}d_a d_b}{I_2}\right)\right]^{-1} \tag{5.33}$$

where

$$d_a = d_2 \sin(\theta_2 + \phi_2), \qquad d_c = d_1 \sin(\theta_1 + \phi_1) \tag{5.4}$$

$$d_b = d_2 \cos(\theta_2 + \phi_2), \qquad d_d = d_1 \cos(\theta_1 + \phi_1) \tag{5.5}$$

and

$$d_e = d_c - \mu d_d, \qquad d_f = d_a - \mu d_b \tag{5.37}$$

## *Energy Loss*

$$T_L = \tfrac{1}{2}\overline{m}(1 + e)qv_m^2\left[2 + 2\mu r - (1 + e) \cdot q\left(1 + \mu^2 + \frac{\overline{m}d_e^2}{I_1} + \frac{\overline{m}d_f^2}{I_2}\right)\right] \tag{5.35}$$

$$v_{rn} = (v_{2n} - d_a\omega_2) - (v_{1n} + d_c\omega_1) \tag{5.36}$$

$$r = \frac{(v_{2t} + d_b\omega_2) - (v_{1t} - d_d\omega_1)}{(v_{2n} - d_a\omega_2) - (v_{1n} + d_c\omega_1)} \tag{5.38}$$

$$\mu_m(e) = \frac{\dfrac{r}{1 + e}A + \dfrac{1}{1 + e}B}{1 + C + \dfrac{r}{1 + e}B - \dfrac{e}{1 + e}\dfrac{B^2}{A}} \tag{5.39}$$

where

$$A = 1 + \frac{\overline{m}d_c^2}{I_1} + \frac{\overline{m}d_a^2}{I_2}, \qquad B = \frac{\overline{m}d_c d_d}{I_1} + \frac{\overline{m}d_a d_b}{I_2},$$

$$C = \frac{\overline{m}d_d^2}{I_1} + \frac{\overline{m}d_b^2}{I_2} \tag{5.40}$$

$$\mu_{\max} = \frac{rA + B}{1 + C + rB} \tag{5.41}$$

$$\mu_0(e) = \frac{\dfrac{r}{1 + e}A + B}{1 + C + \dfrac{r}{1 + e}B} \tag{5.42}$$

$$\mu_T(e) = \frac{rA + eB}{2rB(1 + e)(1 + c)} \cdot \left\{1 \pm \left[1 + (1 - e)\,A\,\frac{2rB + (1 + e)(1 + c)}{(rA + eB)^2}\right]^{1/2}\right\} \tag{5.43}$$

$$|\mu| \leq |\mu_c| = \min[f, |\mu_0|, |\mu_T|] \tag{5.46}$$

and

$$\text{sgn}(\mu) = \text{sgn}(\mu_{\max}) \tag{5.47}$$

## PROBLEMS

**5.1** State the conditions under which the rigid body solution equations reduce to the point-mass solution equations of Chapter 2.

**5.2** Two identical spheres each with a radius $R = 1$ have a direct, perfectly elastic central collision with initial spin, where $v_{1n} = -2$, $v_{2n} = 2$, $v_{1t} = v_{2t} = 0$, $\omega_1 = 1$, and $\omega_2 = -1$. The spheres are such that $m_1 = m_2 = 1$ and $I_1 = I_2 = \frac{2}{5}$. Determine the final velocities if friction coefficients are $f = 0$ and $f = 0.6$.

**5.3** Two identical spheres each with a radius $R = 1$ have a direct, perfectly elastic central collision with initial spin, where $v_{1n} = -2$, $v_{2n} = 2$, $v_{1t} = v_{2t} = 0$, $\omega_1 = 1$, and $\omega_2 = +1$. The spheres are such that $m_1 = m_2 = 1$ and $I_1 = I_2 = \frac{2}{5}$. Determine the final velocities for friction coefficients of $f = 0$ and $f = 0.6$.

**5.4** The rod of example 5.3 is replaced by a rigid rod hinged at mid-length, point $H$, as shown. A stretched spring is connected to points

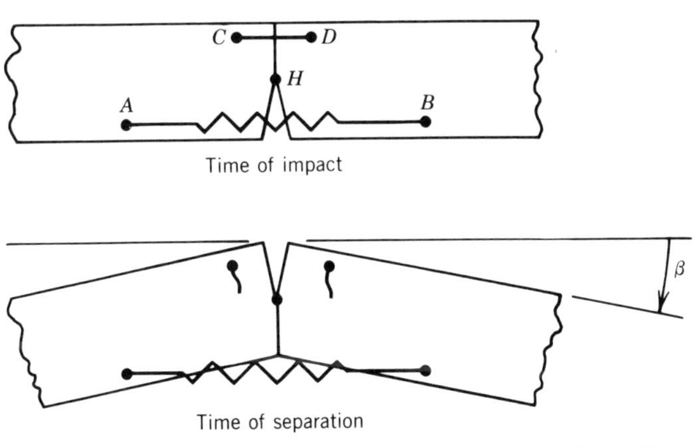


**Figure P5.4.** System configurations at time of contact and time of separation.

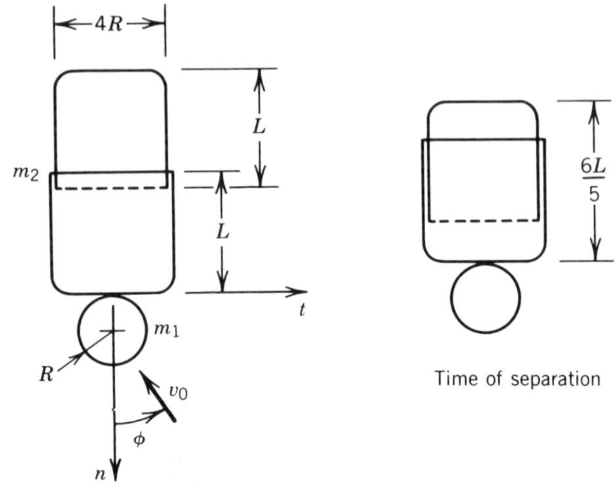


**Figure P5.5.** System configurations at time of contact and time of separation.

*A* and *B* with a fragile thread tying *C* to *D*. Upon impact by the disk, the thread breaks and the spring causes the rod to form the angle $\beta$ prior to separation as shown. Under what conditions does the solution of example 5.3 apply to this problem?

**5.5** Two circular cylindrical cups are held together by friction in a telescoped fashion as shown, with an overall length of $9L/5$. Each cup has a mass of $m = m_2/2$. A solid sphere of mass $m_1 = m$ and velocity $v_0$ strikes the bottom of the initially stationary cups and causes them to permanently compress to the length $6L/5$ prior to separation. Determine the final velocities for a point impact ($e_m = -1$) and arbitrary $e$ and $\mu$.

**5.6** Show that the energy loss $T_L$ from Eq. 5.35 is zero for two values of the impulse ratio $\mu$. For $e_m = -1$ and $e = 1$ show that one of these is equal to $\mu_{\text{max}}$. That is, derive Eq. 5.43.

## REFERENCES

1. Barkan, P., "Impact," *Mechanical Design and Systems Handbook*, Section 16, McGraw-Hill, New York, 1974.
2. Pestel, E. C. and Thomson, W. T., *Dynamics*, McGraw-Hill, New York, 1968.

# CHAPTER 6

# PLANAR BARRIER COLLISIONS

When one of two colliding rigid bodies is significantly more massive than the other, its velocity changes may be negligible. Considerable simplification of the equations occurs. If the momentum change of $m_2$ is assumed to be zero, then the velocity changes of $m_1$ alone are to be found. The number of unknown velocities is reduced to three from six. Masses $m_2$ and $m_1$ are no longer considered to be a system of two masses, and forces and impulses are no longer interior to a system. Consequently, momentum is not conserved for $m_1$. This implies that Eqs. 5.1, 5.2, and 5.9 no longer apply, but Eqs. 5.3, 5.8, and 5.15 still remain valid.

Although the problem of the impact of a single rigid body against a barrier is a special case of the impact of two rigid bodies, it is an important problem with many applications and is therefore covered independently. Another reason for separate coverage is that with a simpler set of equations, additional insight can be gained into the effects of the coefficients $e$, $e_m$, $e_t$, and $\mu$. The simpler form of the equations also permits a convenient introduction of a new coefficient, encountered in the presence of spin and torsional effects. This will fit into the three-dimensional problem covered in Chapter 7.

The format of this chapter is to present first a general set of the system equations. With only three unknowns, they are easily solved. Because of the generality, the solution equations are long and complicated and are not presented. Rather, solution equations for numerous special cases are covered in detail in order to gain insight into the solutions, frequently with the aid of experimental data.

Some specific and important questions will be addressed in this chapter. These are:

1. The effects of external forces are usually neglected. How can they be included? When are they significant?
2. Is dry, or Coulomb, friction an appropriate model for tangential forces during an impact?
3. Under the assumption of dry friction, can the relative tangential (sliding) velocity change sign during the contact interval?
4. In a collision with a rigid barrier, can a rigid body gain energy?

The effects of external forces are viewed in Example 6.1, where it is seen that the forces as well as the duration of contact must be known.

Dry friction can be an acceptable model for tangential forces during some collisions. As a rough rule of thumb, this is true if a relatively soft object collides with a flat, relatively hard surface. Experimental data are examined in order to determine the reasonableness of this friction model.

The question of velocity reversal is one which has been addressed in the literature.[1] In the presence of dry friction and in the absence of tangential restitution, tangential velocity reversals can occur during the collision of rigid bodies. (This is not true for particle collisions, and thus the question arises for the first time in this book.) The velocity reversal is related to the noncentral geometry of the collision; its effects are revealed by reviewing energy loss.

Energy of a single rigid body can increase during a collision; energy of a system of rigid bodies cannot increase simply as a result of a collision. Certainly a baseball gains a significant amount of energy as a result of its collision with a well swung bat. The ball can never gain energy when it falls on the ground or hits any stationary object.

These and many other topics, such as an examination of the moment coefficient $e_m$, are discussed in this chapter.

## 6.1 GENERAL SYSTEM EQUATIONS

Since the velocity changes of only one mass are being sought, the subscripts 1 and 2 are dropped whenever possible from the variables. Figure

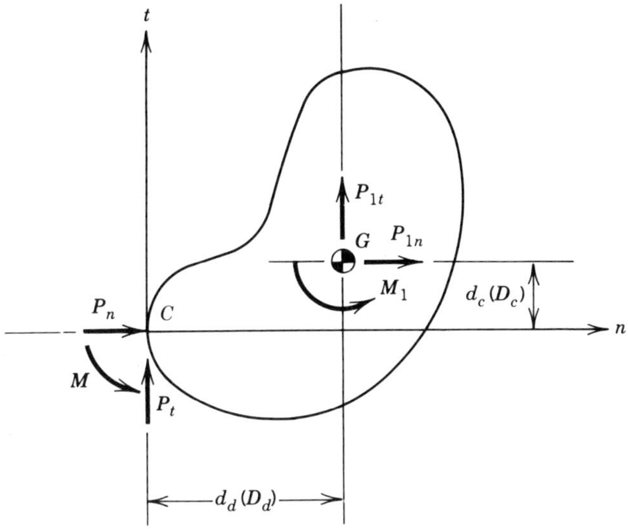


**Figure 6.1.** Diagram and coordinates of rigid body.

6.1 shows a free body diagram of a rigid body striking a massive surface with a common tangent and normal at $C$. Point $G$ is the mass center. Impulses $P_{1n}$, $P_{1t}$, and $M_1$ are externally applied and are considered known; $P_n$, $P_t$, and $M$ are unknown. The distances $d_c$ and $d_d$ are those that correspond to the initial dimensions of the body. If permanent deformations occur during the impact, the corresponding final distances are $D_c$ and $D_d$.

The principle of impulse and momentum along the two coordinate axes provides two equations:

$$P_{1n} + P_n = m(V_n - v_n) \tag{6.1}$$

and

$$P_{1t} + P_t = m(V_t - v_t) \tag{6.2}$$

As in Chapter 5, any permanent dimensional changes are assumed to take place instantaneously when approach ends and rebound begins. Angular impulse and momentum are related by

$$J\Omega - I\omega = \frac{P_n}{1+e}d_c + \frac{eP_n}{1+e}D_c - \frac{P_t}{1+e}d_d - \frac{eP_t}{1+e}D_d + M + M_1 \qquad (6.3)$$

This provides an expression for the unknown moment $M$, which can be written as

$$\frac{M}{I} = \frac{J}{I}\Omega - \omega - \Gamma P_n + \Delta P_t - \frac{M_1}{I} \qquad (6.4)$$

where

$$\Gamma = \frac{d_c + eD_c}{(1+e)I} \quad \text{and} \quad \Delta = \frac{d_d + eD_d}{(1+e)I} \qquad (6.5)$$

As before, the coefficient of restitution $e$ is defined as the negative of the ratio of the final to initial normal relative velocity at point $C$. If the barrier has velocity components $v_{bn}$ and $v_{bt}$, the equation involving $e$ can be written as

$$V_n + D_c\Omega - v_{bn} = -e(v_n + d_c\omega - v_{bn}) \qquad (6.6)$$

Tangential effects can be accounted for using any of three equations. The most general is simply to define the impulse ratio $\mu$ as

$$P_t = \mu P_n \qquad (6.7.1)$$

If it is known that relative tangential motion at $C$ ceases at or prior to separation, then

$$V_t - D_d\Omega - v_{bt} = 0 \qquad (6.7.2)$$

Or, if restitution of tangential velocities occurs, then a tangential restitution coefficient $e_t$ can be defined. Using the form of Eq. 6.6 provides a corresponding equation,

$$V_t - D_d\Omega - v_{bt} = -e_t(v_t - d_d\omega - v_{bt}) \qquad (6.7.3)$$

Experimental results are examined in this chapter illustrating the validity of Eq. 6.7.3 in some instances. As noted in Chapter 5, Eq. 6.7.1 is general enough to permit Eqs. 6.7.2 and 6.7.3 to be considered as special cases.

Another system equation is obtained from the definition of a moment coefficient $e_m$. From Eq. 5.11, this equation is

$$e_m M = (1 + e_m)I\Omega \qquad (6.8)$$

At this point, six unknowns exist. These are $V_n$, $V_t$, $\Omega$, $P_n$, $P_t$, and $M$. To simplify the system equations, the impulses can be eliminated using Eqs. 6.1, 6.2, and 6.4. With the impulses eliminated from Eqs. 6.6, 6.7.1, and 6.8, three system equations result. These are

$$V_n + D_c\Omega - v_{bn} = -e(v_n + d_c\omega - v_{bn}) \qquad (6.6)$$

$$\mu V_n - V_t = \mu v_n - v_t - (P_{1t} - \mu P_{1n})/m \qquad (6.9)$$

and

$$e_m m\Gamma V_n - e_m m\Delta V_t + [1 - e_m(J/I - 1)]\Omega$$
$$= e_m[m\Gamma v_n - m\Delta v_t - \omega + \Gamma P_{1n} - \Delta P_{1t} - M_1/I] \qquad (6.10)$$

Two alternate sets of system equations can be found when Eq. 6.7.2 or 6.7.3 is used in place of Eq. 6.9. Whatever form is used, these system equations can be solved easily for the unknowns $V_n$, $V_t$, and $\Omega$. The expressions are complicated and provide little insight into the processes involved. Moreover, general solution equations provide little convenience, since numerical solutions are more easily adapted to problem solving. For these reasons, general solution equations are not presented and can be found by the reader if needed.

Once final velocities are found, the unknown impulses can be calculated from the earlier equations.

As discussed in earlier chapters, an inverse problem (as well as mixed problems) can be formulated where the final velocity components are known and the initial velocities are unknown. Again, caution must be used since the value of $e = 0$ provides a singular set of equations (see Eq. 6.5). If used, a zero value of the tangential coefficient $e_t$ can also cause difficulties in the inverse problem. This applies to $e_m$ as well. The remainder of this chapter is devoted to the solution and examination of several special cases of these system equations.

■ ***Example 6.1***

PROBLEM: An inelastic solid sphere ($0 < e < 1$) with radius $R$ and radius of gyration $k$ collides with a massive stationary surface. Assume no permanent deformation, and no moment impulse. Use arbitrary initial velocity components $v_n$ and $v_t$; let $\omega = 0$. If the duration of contact is $\tau$, determine the final velocities for a collision with a horizontal surface and the same for a vertical surface. Determine the impulse ratio $\mu_0$ that discriminates sliding from rolling at separation.

*Solution:* The effect of gravity will be treated as a constant vertical force acting over the duration $\tau$, thus creating known external impulse components. The solution equations can be found from system equations 6.6, 6.9, and 6.10. The solution equations are

$$V_n = -ev_n \tag{a}$$

$$V_t = v_t - \mu(1 + e)v_n + (P_{1t} - \mu P_{1n})/m \tag{b}$$

$$\Omega = \mu \frac{R}{k^2}\left[(1 + e)v_n + \frac{P_{1n}}{m}\right] \tag{c}$$

When sliding ends at or prior to separation, $V_t - R\Omega = 0$. Substitution of $V_t$ and $\Omega$ from Eqs. b and c into this expression provides the impulse ratio $\mu_0$, where

$$\mu_0 = \frac{v_t + P_{1t}/m}{(1 + R^2/k^2)[(1 + e)v_n + P_{1n}/m]} \tag{d}$$

This value of $\mu$ can be substituted into Eqs. b and c, which gives

$$V_t = R\Omega = \frac{v_t + P_{1t}/m}{1 + k^2/R^2} \tag{e}$$

### Case I: Horizontal Surface

In this case, $P_{1n} = -mg\tau$ and $P_{1t} = 0$. Equations b and c become

$$V_t = v_t - \mu(1 + e)v_n + \mu g\tau \tag{f}$$

$$\Omega = \mu \frac{R}{k^2}[(1 + e)v_n - g\tau] \tag{g}$$

If sliding ends, the sphere will be rolling at separation. Equation e gives

$$V_t = R\Omega = \frac{v_t}{1 + k^2/R^2} \tag{h}$$

### Case II: Vertical Surface

Here, $P_{1n} = 0$ and $P_{1t} = -mg\tau$, which give

$$V_t = v_t - \mu(1 + e)v_n - g\tau \tag{i}$$

and

$$\Omega = \mu \frac{R}{k^2}(1 + e)v_n \tag{j}$$

For rolling at separation

$$V_t = R\Omega = \frac{v_t - g\tau}{1 + k^2/R^2} \tag{k}$$

*Comments:* Note that the sliding solution—Eqs. a, b, and c—contains both external impulse components $P_{1t}$ and $P_{1n}$. The normal component $P_{1n}$ does not appear in the rolling solution, Eq. e. However, if the tangential effect is caused by dry friction, $\mu = f$, and Eq. d shows that a lower value of friction coefficient can result in rolling motion in the presence of gravity than when gravity is not included.

Also note that the significance of an external impulse can be judged by a comparison of the $g\tau$ term with the initial tangential velocity (see Eq. k). If the duration $\tau$ is small and the initial velocity $v_t$ is large, the external impulse can be neglected. ■

## ■ *Example 6.2*

PROBLEM: An automobile is propelled at a speed of 30 mph into a fixed vertical barrier as shown. It contacts the barrier over a surface with the normal impulse acting at height $h$ above the floor. Measured values of the velocities of the mass center and the angular velocities are

$$V_n = 3.44 \text{ ft/s}, \qquad v_n = -44.0 \text{ ft/s}$$
$$V_t = 1.80 \text{ ft/s}, \qquad v_t = 0$$
$$\Omega = 0.430 \text{ rad/s}, \qquad \omega = 0$$

Assume no sliding at separation along the vertical barrier surface and that the pitch moment of inertia does not change significantly due to deformation. Determine $e$, $\mu$, the height $h$, and $P_n$. The following information is known about the vehicle:

Weight = 2900 lb
Pitch moment of inertia = 522 ft-lb-s$^2$
Height of mass center, $h_G$ = 2 ft
Undeformed distance, $d_d$ = 5.4 ft
Deformed distance, $D_d$ = 4.2 ft

*Solution:* It is assumed that the barrier surface is uniform and so $e_m = -1$. The system equations must be solved for the conditions that $v_t =$

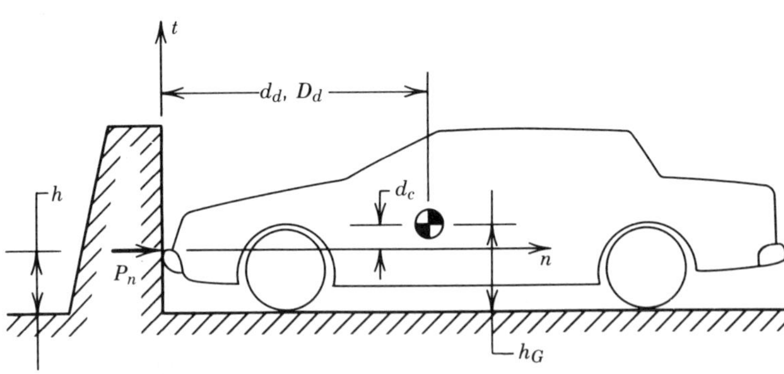


**Figure E6.2.** Frontal impact with a rigid barrier.

$\omega = 0$ and $d_c = D_c$ but $d_d \neq D_d$. With no external impulses, Eqs. 6.6, 6.9, and 6.10 have the solution

$$V_n = v_n - q(1 + e)v_n \tag{a}$$

$$V_t = -\mu q(1 + e)v_n \tag{b}$$

and

$$\Omega = -q(1 + e)v_n \frac{d_c(1 - \mu\eta)}{k^2} \tag{c}$$

where

$$\eta = \frac{d_d + eD_d}{(1 + e)d_c} \tag{d}$$

and

$$q = \left[1 + (1 - \mu\eta)\,(d_c^2/k^2)\right]^{-1} \tag{e}$$

These equations are nonlinear for the unknowns $e$, $\mu$, and $d_c$. An iterative solution is used. In sequence, Eqs. a, b, and c are solved for $e$, $\mu$, and $d_c$. Iteration is begun with the value $q = 1$. If the car's initial speed is $v_0$, then $v_n = -v_0$,

$$e = \frac{1}{q}\left(1 - \frac{V_n}{v_0}\right) - 1 = 0.07818 \tag{f}$$

$$\mu = \frac{V_t}{(1 + e)qv_0} = 0.03794 \tag{g}$$

and

$$d_c = \mu_0 \frac{d_d + eD_d}{1 + e} + \frac{k^2\Omega}{(1 + e)qv_0} = 0.20159 \tag{h}$$

A value of $q = 0.99288$ is obtained from the current values of the unknowns. With this value of $q$, Eqs. f, g, and h give (second iteration)

$$e = 0.086$$

$$\mu = 0.038$$

$$d_c = 0.243 \text{ ft}$$

These provide the same value of $q$, and so the iteration process is completed. The normal impulse $P_n$ is given by $m(V_n - v_n) = 4273$ lb-s and $h = 1.757$ ft. ■

## 6.2 TIP COLLISION OF A SLENDER ROD

A thorough solution and discussion of the collision of the tip of a long slender rod with a massive barrier is presented in this section. Before proceeding to the mathematics of the problem, it is beneficial to discuss the different physical conditions that the tip of the rod can encounter. Figure 6.2*a* shows the type of sketch typically drawn to illustrate this problem. Actual conditions at the impact region at $C$ can include a reasonably flat hard surface with friction such as shown in Fig. 6.2*b* or those that permit significant penetration such as Fig. 6.2*c*. The latter would correspond for example to a javelin "sticking" into the ground and would be represented by the coefficient values of $e = 0$, $e_m = 0$, and $\mu = \mu_0$. If the surface has significant irregularities, such as shown in Fig. 6.2*d*, then it may be necessary to know the local tangent and normal, as indicated by the coordinates $n'$ and $t'$. All cases can be covered as special cases of diagram 6.2*a* and an appropriate choice of coefficients.

To simplify the equations, let the external impulses be negligible, the rod's dimensions be constant, and the barrier be fixed. Under these conditions the solution to Eqs. 6.6, 6.9, and 6.10 can be written as

$$V_n = v_n - q_m[(1 + e)(v_n + d_c\omega) - (1 - e_m)d_c\omega] \quad (6.11)$$

$$V_t = v_t - \mu q_m[(1 + e)(v_n + d_c\omega) - (1 + e_m)d_c\omega] \quad (6.12)$$

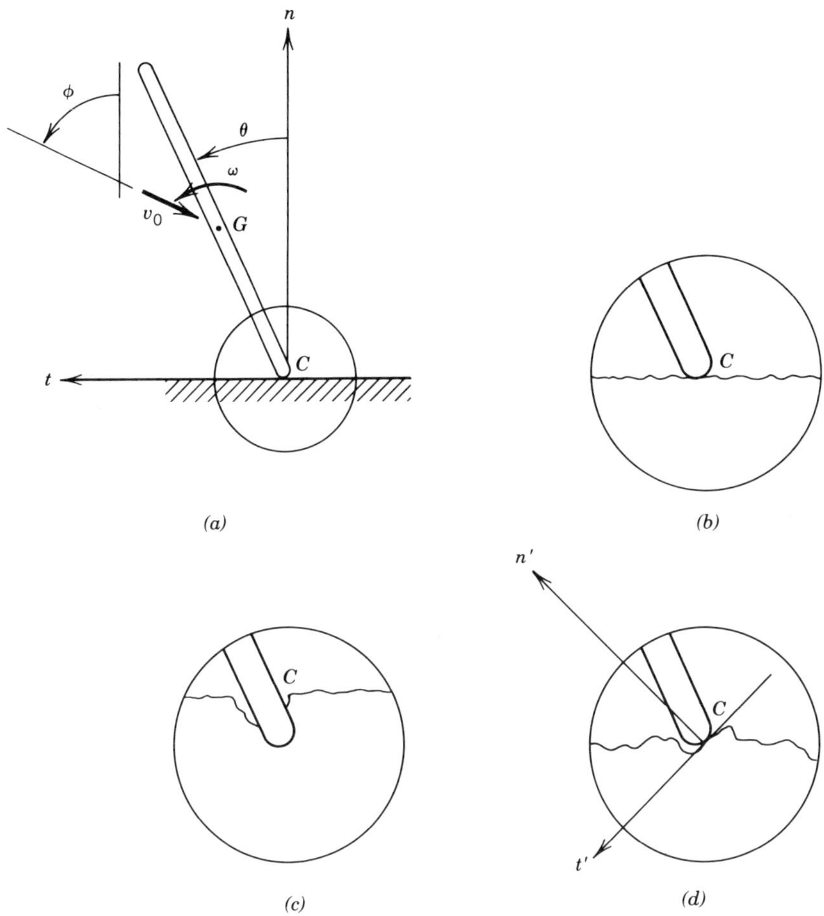


**Figure 6.2.** Various tip conditions, slender rod impact: (*a*) idealized case; (*b*) hard, rough surface; (*c*) soft penetrable surface; (*d*) significant surface irregularities.

and

$$\Omega = \omega + q_m\left[me_m(1 + e)(d_c - \mu d_d)(v_n + d_c\omega)/I - (1 + e_m)\omega\right] \tag{6.13}$$

where

$$q_m = \left[1 - md_ce_m(d_c - \mu d_d)/I\right]^{-1} \tag{6.14}$$

It can be noted that these equations are valid for any rigid body, since they have not been specialized to the rod problem. If the collision is not accompanied by a moment impulse at the tip, that is, $M = 0$ (and $e_m = -1$), the energy loss of the collision is given by

$$T_L = \tfrac{1}{2}m(v_n + d_c\omega)^2(1 + e)q\{2 + 2\mu r - (1 + e) \cdot q[1 + \mu^2 + (d_c - \mu d_d)^2/k^2]\} \tag{6.15}$$

where

$$q = [1 + d_c(d_c - \mu d_d)/k^2]^{-1} \tag{6.16}$$

and $I = I_G = mk^2$, where $k$ is the radius of gyration. Again, this equation is for $e_m = -1$, but for any rigid body.

Attempts have been made[2-4] to solve various special cases of this problem, such as with no friction. They have been only partly successful and differ in various ways from the above solution.

With the solution equations available, various initial conditions can be examined. The initial conditions and parameters chosen are listed in Table 6.1.

Figures 6.3 through 6.6 show the energy loss plotted as a function of $\mu$ for a range of values of the coefficient $e$, each figure corresponding to a different value of the initial tangential velocity $v_t$. For these illustrations, the energy loss is normalized to the initial kinetic energy of the rod. Tables 6.2–6.5 list some of the final velocity values corresponding

**TABLE 6.1 Slender Rod Impact Example; Physical Parameters**

| | |
|---|---|
| Mass | $m = 1$ |
| Rod length | $L = 1$ |
| Moment of inertia | $I = mk^2$ |
| Radius of gyration | $k = L/\sqrt{12}$ |
| Angle | $\theta = 45°$ |
| Centroidal distances | $d_c = d_d = L/2\sqrt{2}$ |
| Initial velocities | $v_n = -v_0 \cos\phi = -1$ |
| | $v_t$ (varied) |
| | $\omega = 0$ |

to these figures. Table 6.6 shows some of the important values of the impulse ratio, $\mu$.

This problem is of interest because its kinematics are relatively simple to visualize, yet contain enough complexity to permit nonintuitive results. One question that arises concerns the possibility of a reversal in sign (during the duration of contact) of the tangential velocity at point $C$. Others are the previously recognized question of an appropriate value of $\mu$ in the case of dry friction and the related question of conservation of energy. The discussion that follows takes the point of view that dry friction exists between the rod and surface.

### 6.2.1 Discussion of Fig. 6.3, $v_t = 0$

Figure 6.3 shows energy loss for a rod falling vertically with no initial centroidal tangential velocity $v_t$. Following the collision, a counterclockwise angular velocity exists, since point $C$ is either brought to rest or rebounds with a positive vertical velocity component, owing to the kinematical constraint $V_{cn} = -ev_{cn}$. Because of the angular velocity, point $C$ develops a negative final tangential velocity component $V_{ct}$. This dem-

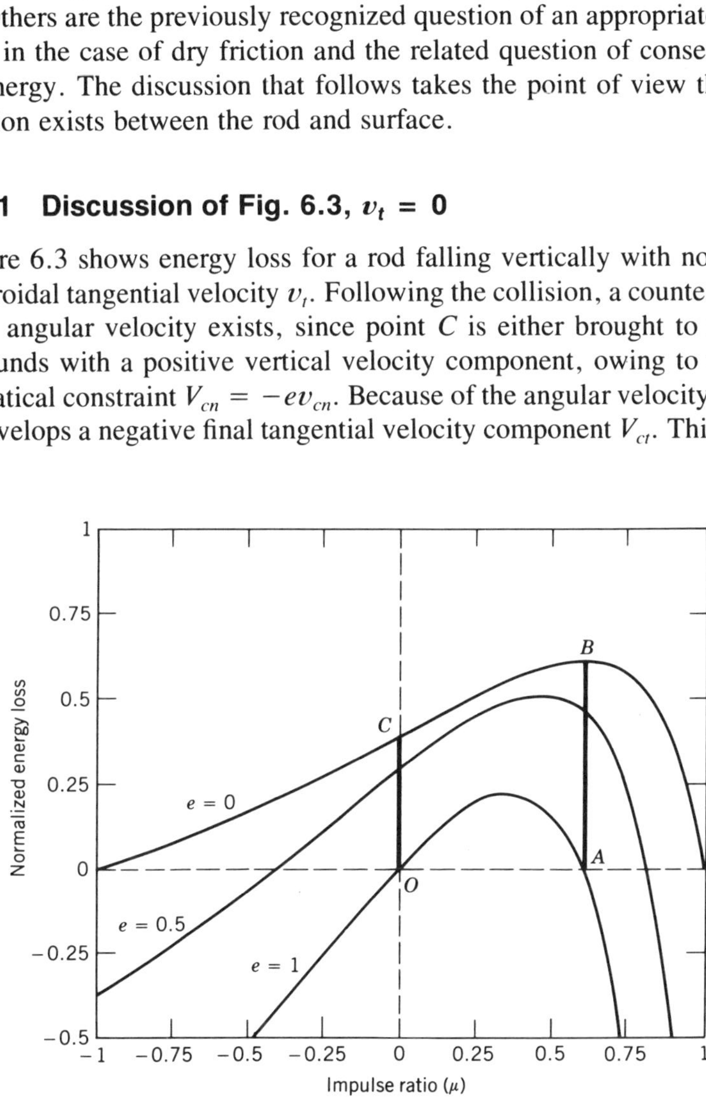


**Figure 6.3.** Energy loss for falling rod impact, $v_t = 0.0$.

**TABLE 6.2 Slender Rod Impact Final Velocities for $v_t = 0.0$ m/s**

| $e$ | $\mu$ | $V_n$ | $V_t$ | $\Omega$ | $V_{cn}$ | $V_{ct}$ |
|---|---|---|---|---|---|---|
| 0 | 0.00 | −0.600 | 0.000 | 1.697 | 0.000 | −0.600 |
| 0 | 0.60 | −0.375 | 0.375 | 1.061 | 0.000 | 0.000 |
| 0.5 | 0.00 | −0.400 | 0.000 | 2.546 | 0.500 | −0.900 |
| 0.5 | 0.45 | −0.178 | 0.370 | 1.918 | 0.500 | −0.308 |
| 0.5 | 0.60 | −0.062 | 0.563 | 1.591 | 0.500 | 0.000 |
| 1.0 | 0.00 | −0.200 | 0.000 | 3.394 | 1.000 | −1.200 |
| 1.0 | 0.35 | 0.013 | 0.354 | 2.793 | 1.000 | −0.633 |
| 1.0 | 0.60 | 0.250 | 0.750 | 2.121 | 1.000 | 0.000 |

onstrates that upon impact a rigid body with an initial tangential velocity at the contact point $v_{ct} = 0$ can develop a nonzero final value $V_{ct}$. This is due to the kinematics (creation of a rotational velocity $\Omega$ owing to the tip's normal velocity change) not due to a tangential impulse. Owing to the nature of friction, several points can be made. The effect of a tangential frictional impulse can only reduce this tangential velocity. Since $P_n > 0$ and any value of $P_t$ that develops due to friction must also be positive (opposite $V_{ct}$), $\mu$ cannot be negative. Consequently, for dry friction, valid values of $\mu$ must correspond to the region to the right of line *OC* in Fig. 6.3. Friction at the contact point can only decrease the tip velocity gained in the frictionless case. Furthermore, at most friction can prevent the tip from gaining any tangential tip velocity. This means that $\mu \leq \mu_0$. For these initial conditions ($v_t = 0$), $\mu_0$ can be found using Eq. 5.44; $\mu_0 = 0.6$. It is independent of $e$ and shown as the vertical line *AB* in Fig. 6.3. Consequently, $0 \leq \mu < \mu_0$. Note that, coincidentally, $\mu_{\max} = \mu_0$. For this collision the energy loss is bounded by the energy-loss curves O-A and B-C and by the vertical lines O-C and A-B.

### 6.2.2 Discussion of Fig. 6.4, $v_t = 0.2$

Figure 6.4 shows the energy loss when a small, positive initial tangential velocity exists; here $v_t = 0.2$. Several interesting observations can be made. As in the previous discussion, the constraint $V_{cn} = -ev_{cn}$ creates the condition for a negative $V_{ct}$. The small initial velocity $v_{ct} = v_t = +0.2$ helps the friction reduce the magnitude of $V_{ct}$ compared to the previous case. Consequently, during the interval of contact a reversal in sign occurs for the tangential contact velocity (and the tangential contact

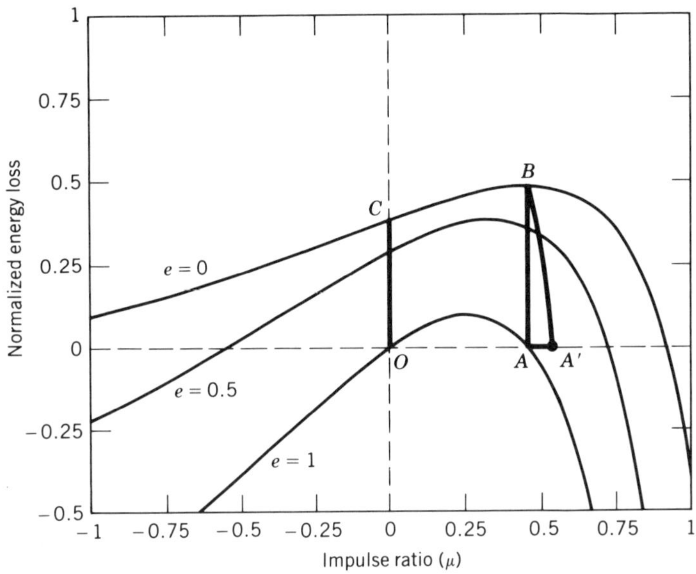


**Figure 6.4.** Energy loss for falling rod impact, $v_t = 0.2$.

force). Table 6.3 shows that all values of $V_{ct}$ are zero or negative. Again, $P_n > 0$ and $P_t \geq 0$, so $\mu > 0$; hence, the bounding line $OC$ in Fig. 6.4 as in the previous example. Note, however, that $\mu_0$ from Eq. 5.44 now becomes $\mu_0 = (10 + 15e)/(22 + 25e)$. These values are indicated on the energy loss curves of Fig. 6.4 by the curve from point $A'$ to $B$ and represents a bound on $\mu$ because friction can reduce $V_{ct}$ to zero but cannot cause it to become positive. But the presence of friction cannot re-

**TABLE 6.3 Slender Rod Impact Final Velocities for $v_t$ = 0.2 m/s**

| $e$ | $\mu$ | $V_n$ | $V_t$ | $\Omega$ | $V_{cn}$ | $V_{ct}$ |
|---|---|---|---|---|---|---|
| 0 | 0.00 | −0.600 | 0.200 | 1.697 | 0.000 | −0.400 |
| 0 | 0.45 | −0.450 | 0.450 | 1.275 | 0.000 | 0.000 |
| 0.5 | 0.00 | −0.400 | 0.200 | 2.546 | 0.500 | −0.700 |
| 0.5 | 0.35 | −0.241 | 0.466 | 2.094 | 0.500 | −0.275 |
| 0.5 | 0.45 | −0.178 | 0.570 | 1.918 | 0.500 | −0.108 |
| 1.0 | 0.00 | −0.200 | 0.200 | 3.394 | 1.000 | −1.000 |
| 1.0 | 0.25 | −0.059 | 0.435 | 2.995 | 1.000 | −0.624 |
| 1.0 | 0.45 | 0.096 | 0.693 | 2.557 | 1.000 | −0.211 |

duce $V_{ct}$ to zero for all values of $e$, particularly high values near 1. To do so would require an increase in the rod's energy. The condition that $T_L \geq 0$, corresponding to $\mu_c = \mu_T$, is illustrated in Fig. 6.4 by the line from $A$ to $A'$. Consequently, realistic values of the impulse ratio must be zero or positive and are bounded below by the line $OC$. Realistic values of $\mu$ are bounded above by the values $\mu_T$ and $\mu_0$, illustrated in Fig. 6.4 by the curves $AA'B$.

In the above cases, the vertical line $OC$ represents more than impact against a frictionless surface, $f = 0$. When tangential velocity reversals occur, the frictional force changes sign and a value of $\mu = 0$ for any value of $f$ can reflect a cancellation of negative and positive portions of the tangential impulse. For example, this is true for point $A$ of Fig. 6.5, where point $A$ and its value $\mu_T = 0$ provide the solution for all cases of elastic impact, $e = 1$, regardless of the value of the friction coefficient.

### 6.2.3 Discussion of Fig. 6.5, $v_t = 0.6$

In this case, the rod is given an initial tangential velocity equal to the value of $V_{ct}$ when $v_t = 0$, and $e = 0$ and $\mu = 0$. Table 6.4 shows that

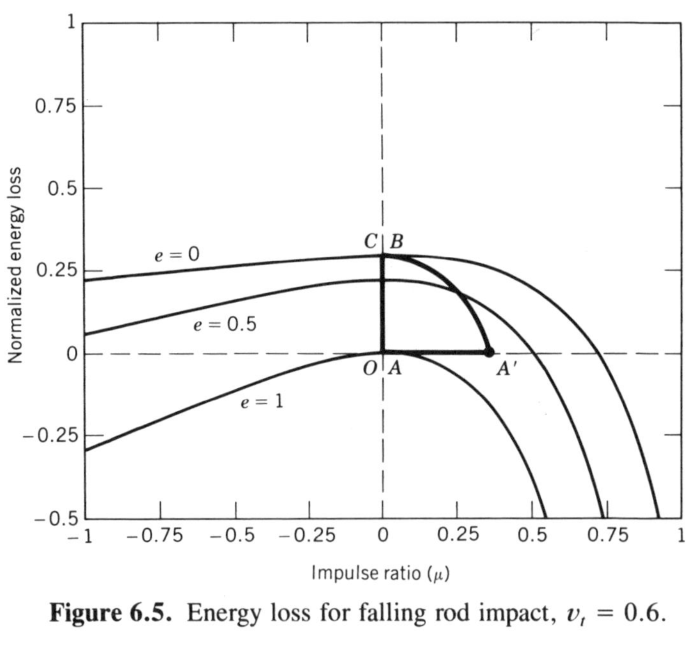


**Figure 6.5.** Energy loss for falling rod impact, $v_t = 0.6$.

**TABLE 6.4 Slender Rod Impact Final Velocities for $v_t = 0.6$ m/s**

| $e$ | $\mu$ | $V_n$ | $V_t$ | $\Omega$ | $V_{cn}$ | $V_{ct}$ |
|---|---|---|---|---|---|---|
| 0 | 0 | −0.600 | 0.600 | 1.697 | 0.000 | 0.000 |
| 0.5 | 0.000 | −0.400 | 0.600 | 2.546 | 0.500 | −0.300 |
| 0.5 | 0.251 | −0.294 | 0.776 | 2.246 | 0.500 | −0.018 |
| 1.0 | 0.000 | −0.200 | 0.600 | 3.394 | 1.000 | −0.600 |

with $v_t = 0.6$ and $e = 0$, $V_{ct} = 0$ and must correspond to $\mu_{\max}$. That is, the initial velocity counteracts the effects noted above which make $V_{ct}$ negative. For larger values of $e$, $V_{ct}$ is still negative but lower in magnitude than earlier cases. As above, friction can reduce $V_{ct}$ to zero for some values of $e$; see curve $A'B$ in Fig. 6.5. The values $\mu_T$ which correspond to $T_L = 0$ complete the bound on $\mu$ and are illustrated by the line $AA'$.

Again, $\mu$ cannot be negative, and so the curves $OC$ and $AA'B$ serve to bound $\mu$ and furnish the critical values for this set of initial conditions. Note here that $\mu_{\max} = 0$, so that points $B$ and $C$ coincide. Also, for $e = 1$, $\mu_T = 0$ so that points $A$ and $O$ coincide.

### 6.2.4 Discussion of Fig. 6.6, $v_t = -1$

In this case the rod moves initially in such a way that the velocity vector lies along the rod's axes. Here $v_{ct}$ is negative, the kinematical restitution constraint tends to make it more negative (see Table 6.5) and, again, friction can act to reduce the magnitude of $V_{ct}$. From Eq. 5.44, the impulse ratio that causes $V_{ct} = 0$ is $\mu_0 = (8 + 3e)/(8 + 5e)$. This is illustrated in Fig. 6.6 as the curve $A'B$, where $B$ is again the point of maximum energy loss. Here, for $e = 1$, $\mu_0$ is less than $\mu_T$, so that the values of $\mu_0 \leq \mu_{\max}$, a trend not seen earlier.

Note also that the energy loss curves (for different values of $e$) intersect, another trend not seen earlier. This means that for some values of $\mu$, increasing the coefficient of restitution $e$ (making the impact more "elastic") causes more energy to be lost. This is perhaps counterintuitive but stems from the fact that a larger $e$ means a larger $P_n$. A larger $P_n$ with $P_t = \mu P_n$ means that $P_t$ is large and has the effect of increasing the frictional energy loss. Friction seems to lose energy more efficiently here than elasticity can conserve it. Table 6.6 summarizes some of the important values of the impulse ratio, $\mu$.

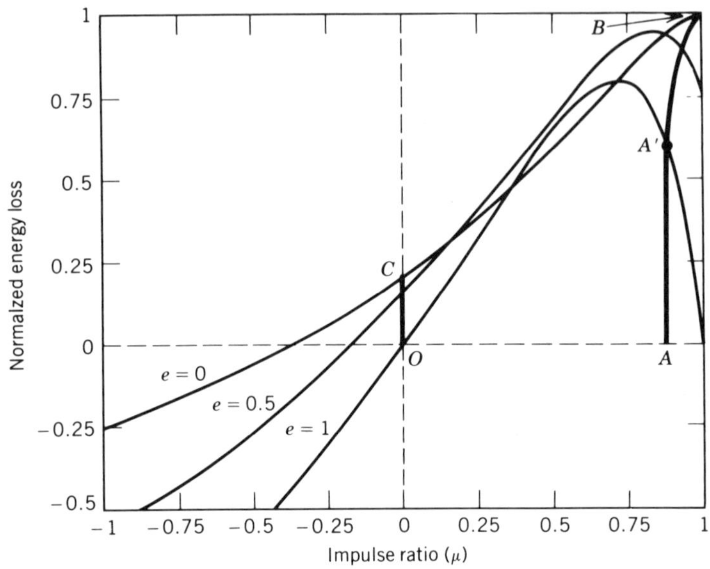


**Figure 6.6.** Energy loss for falling rod impact, $v_t = -1.0$.

**TABLE 6.5 Slender Rod Impact Final Velocities for $v_t = -1.0$ m/s**

| $e$ | $\mu$ | $V_n$ | $V_t$ | $\Omega$ | $V_{cn}$ | $V_{ct}$ |
|---|---|---|---|---|---|---|
| 0 | 0.00 | −0.600 | −1.000 | 1.697 | 0.000 | −1.600 |
| 0 | 1.00 | 0.000 | 0.000 | 0.000 | 0.000 | 0.000 |
| 0.5 | 0.00 | −0.400 | −1.000 | 2.546 | 0.500 | −1.900 |
| 0.5 | 0.85 | 0.224 | 0.041 | 0.729 | 0.500 | −0.235 |
| 0.5 | 1.00 | 0.500 | 0.500 | 0.000 | 0.500 | 0.500 |
| 1.0 | 0.00 | −0.200 | −1.000 | 3.394 | 1.000 | −2.200 |
| 1.0 | 0.70 | 0.379 | −0.034 | 1.756 | 1.000 | −0.655 |
| 1.0 | 1.00 | 1.000 | 1.000 | 0.000 | 1.000 | 1.000 |

**TABLE 6.6 Slender Rod Impact Example Important Impulse Ratio Values**

| | | $\mu_m$ | | | $\mu_0$ | | |
|---|---|---|---|---|---|---|---|
| $v_t$ | $\mu_{max}$ | $e = 0.0$ | $e = 0.5$ | $e = 1.0$ | $e = 0.0$ | $e = 0.5$ | $e = 1.0$ |
| 0.0 | 0.60 | 0.60 | 0.46 | 0.37 | 0.60 | 0.60 | 0.60 |
| 0.2 | 0.46 | 0.46 | 0.33 | 0.26 | 0.46 | 0.51 | 0.53 |
| 0.6 | 0.00 | 0.00 | 0.00 | 0.00 | 0.00 | 0.26 | 0.37 |
| −1.0 | 1.00 | 1.00 | 0.83 | 0.71 | 1.00 | 0.91 | 0.85 |

## 6.3 PARTICLE-SURFACE COLLISIONS: EROSION AND WEAR

If, while colliding with a massive surface, a rigid body has its mass center on or very near to the normal axis (drawn through the contact point), the constant $B$ of Eq. 5.41 is zero. The corresponding constant $C$ becomes the square of the ratio of the body's radius to its radius of gyration. Let this ratio be written here as $\lambda = C = R^2/k^2$. For simplicity, consider the case where the initial angular velocity is zero. Then $\omega = 0$ and $r = v_t/v_n = (\tan \alpha)^{-1}$ is the angle of approach* of the rigid body to the surface. Under these conditions, a normalized energy loss expression is found to be

$$T_L^* = (1 - e^2) \sin^2 \alpha + \frac{1}{1 + \lambda} \frac{\mu}{\mu_c} \left(2 - \frac{\mu}{\mu_c}\right) \cos^2 \alpha \quad (6.17)$$

This quantity is the energy loss $T_L$ divided by the initial kinetic energy of the rigid body. Note, for example, that $\lambda = \frac{5}{2}$ for a solid sphere. This expression will be examined in detail in many of the remaining sections of this chapter. At this point, it is important to note that the first term depends on the coefficient of restitution and the second term depends on the impulse ratio. As the angle of incidence changes, the first term is governed by the normal forces and stresses and the second term depends on the process of generation of the tangential forces.

For this case, it can be shown that $\mu_0 < \mu_T$ and the critical value of $\mu$ is determined by the friction coefficient if appropriate, or $\mu_0 = r/(1 + e)(1 + \lambda)$, whichever is smaller in magnitude. If the rotational energy of the rigid body is negligible, both before and after the collision, then $\lambda$ can be neglected and the rigid body can be treated as a point mass.

### 6.3.1 Coulomb Friction during Impact

When contacting objects slide or tend to slide relative to each other, forces are generated that are generally tangential to the surface of con-

*If $\omega \neq 0$, then $\tan \alpha = (v_t - R\omega)/v_n$ and $\alpha$ is the angle of approach of the contact point rather than the mass center of the rigid body. The above discussion is still meaningful and $\alpha$ simply has a broader interpretation.

tact. These forces are typically referred to as friction. Friction is a surface force, but its resultant along a line of action through a given point is used in almost all analyses. Friction is ubiquitous and enigmatic. A comprehensive, quantitative mechanical model of frictional force generation does not exist. Friction is often attributed to shearing of microscopic asperities of each object. Recent studies[6] show that a few layers of lubricants with thicknesses of the order of molecular dimensions can significantly affect the frictional forces. Frictional forces can depend on many parameters such as material, temperature, surface cleanliness, speed of the sliding objects, rigidity of the materials, kinematics (rolling, sliding, or combinations), and so forth. The most fundamental and simplest model of friction between dry surfaces is that of Coulomb.[5] With a Coulomb model, the sliding friction force between two rigid bodies is proportional to the normal force. The constant of proportionality is the friction coefficient $f$. It is independent of the area of contact. Ideally, it is also independent of relative speed, but, in practice, this is not always true. Because of the complexities of the topic of friction, empirical coefficients are typically used in engineering applications along with a model such as that of Coulomb or one closely related.

The presence of large forces and short duration contact during impact adds another level of complexity to the generation of frictional forces. The question of whether the Coulomb model of friction is applicable to impact has often been raised. Intuition seems to indicate that in some cases the answer should be yes, in other cases, no. In all cases, care must be used in interpreting the impulse ratio $\mu$ and deciding whether or not $\mu$ represents the friction coefficient $f$ in the impact equations. Some experimental data are presented here that show that a friction coefficient $f$ can be used, but it may not always remain constant; it may depend on relative speed. The subject of how experimental data should be viewed in order to determine the nature of friction and the impulse ratio $\mu$ will be covered.

One would expect that under the circumstances of a relatively soft object (such as a basketball) colliding with a flat, rigid surface (such as a wooden floor) that a Coulomb friction model would provide reasonably good results. On the other hand, when a hard sphere (such as a steel ball) collides with a soft surface (such as an elastomer), significant indentation takes place and the contact pressure distribution over the sphere's surface forms part of the tangential force and impulse. Here, a Coulomb model where $f$ is simply used in place of $\mu$ would not be expected to be a good model. When a significant amount of irregular surface distortion occurs, Coulomb friction alone is probably not a good model. Some experimental data indicative of these conditions are presented and discussed in the following sections.

According to the Coulomb model for dry friction, two friction coefficients exist, static and dynamic. By definition of an impact, the colliding objects must have an initial relative velocity. At the very instant contact begins, the normal force is zero and begins to increase. Consequently, the tangential force, which is proportional to the normal force, begins at zero and increases; some sliding always takes place. Under these conditions the appropriate friction coefficient is the dynamic one. As seen in earlier chapters, it is possible that the relative tangential velocity can change sign during rigid body collisions owing to kinematical conditions. In these cases, the velocity must pass through zero and the static friction coefficient may momentarily be appropriate. This is typically ignored, since the static friction coefficient nearly equals the dynamic coefficient for short stopping durations.[8]

Consider a hypothetical experiment where a sphere with an initial velocity $v_0$ impacts a surface and where $v_0$ is held constant for various angles $\alpha$ from 0° to 90°. Suppose the friction coefficient is such that for low values of $\alpha$, sliding continues throughout the contact duration. Since $v_0$ is held constant, the normal impulse increases as $\alpha$ becomes larger. At some angle, say $\alpha_c$, the tangential impulse will become large enough to cause sliding to end at the time of separation. Based on the laws of Coulomb friction, for $\alpha < \alpha_c$, sliding will take place throughout the contact duration and the impulse ratio $\mu$ will equal the dynamic friction coefficient $f$. For $\alpha > \alpha_c$, sliding will end prior to separation, the impulse ratio will equal $\mu_c(\alpha)$, and the sphere will be rolling at separation (refer to Fig. 3.3). For the case of a sphere with $\omega = 0$,

$$\mu_c = \frac{r}{(1 + e)(1 + \lambda)} = \frac{1}{(1 + e)(1 + \lambda) \tan \alpha} \tag{6.18}$$

where $r = v_t/v_n = 1/\tan \alpha$. With allowance for experimental variations, the data from this hypothetical experiment should follow the trends illustrated by the cross-hatched region in Fig. 6.7.

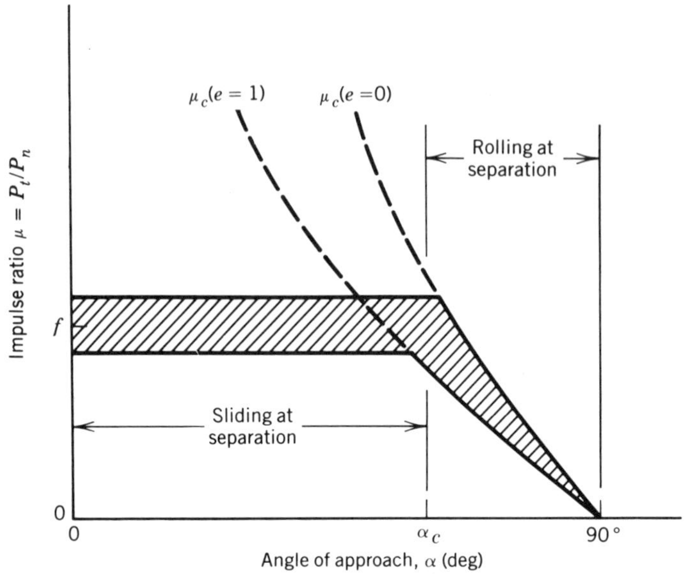


**Figure 6.7.** Variation of impulse ratio with angle of approach (angle of incidence).

The experimental data for $\alpha < \alpha_c$ may not necessarily remain horizontal, since the dynamic friction coefficient may depend on the tangential velocity $v_t$, that is, $f = f(v_t)$. If the friction coefficient decreases with increasing $v_t$, for example, the data points will follow an increasing trend with $\alpha$ for $\alpha < \alpha_c$. If $f$ increases with increasing $v_t$, then an opposite trend will develop.

Note that the data for $\alpha < \alpha_c$ is controlled primarily by the friction coefficient. For $\alpha > \alpha_c$, the angle of incidence controls the trend of the data, as indicated in Eq. 6.18.

Figure 6.8 shows experimental data[7] corresponding to this type of experiment. These data appear to show that a Coulomb friction model is appropriate in some cases. Unfortunately, this was not recognized by the experimentalists and so no measurements were made of the corresponding magnitudes of the nonimpact values of $f$ for the same materials. It should be pointed out that the upper and lower curves in Fig. 6.8 are those of $\mu_c$ for $e = 1$ and $e = 0$. Furthermore, they correspond to the values for $\lambda = 0$. This implies that angular velocities (and therefore rotational energy) are small and that the spheres can be treated as point masses for these experiments.

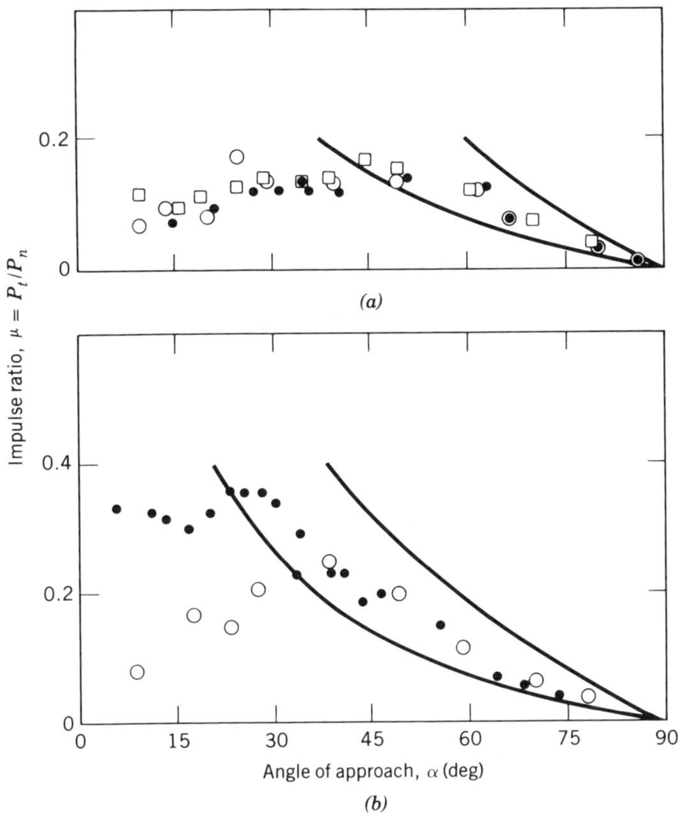


**Figure 6.8.** Experimental impulse ratios (from Ratner and Styller,[7] with permission of Elsevier Sequoia S. A., copyright 1981).

In summary, if a Coulomb or dry friction model is appropriate for impact, then experimental data should follow a trend as indicated in Fig. 6.7. Other trends or a lack of uniformity of data can indicate a tangential process other than simple friction as seen in the next section.

### 6.3.2 Impact with Surface Indentation

When a hard object impacts a massive but relatively soft material, large deformations can occur in the surface. Depending on the object's energy and the material's elasticity and strength, the deformation can be temporary (elastic or viscoelastic) or permanent (plastic). Material may also be removed, leading to erosion of the surface following single or multiple impacts. In some engineering processes, this is done intentionally as a surface treatment, such as shot peening and sand blasting. In other cases, it is an undesirable process such as wind blown sand and ingestion of particles into gas turbine and jet engines. A large body of literature exists on these problems, involving both theoretical and experimental works and is usually referred to as the field of erosion and wear. See, for example, Refs. 9–13. The emphasis in most of these works is the modeling and prediction of the erosion and wear processes from the cumulative effect of multiple impacts either by experiments or by the study of localized deformations, surface pressures, and stresses. Particle dynamics is important, however, since much of the experimental work involves measurement of the particle velocity components both before and after impact. Energy loss is also important; erosion and wear are correlated directly with the particles' kinetic energy loss.[9-13]

Wear is separated into two categories or types. When the particles are blunt or rounded, the wear is primarily due to indentation; this is referred

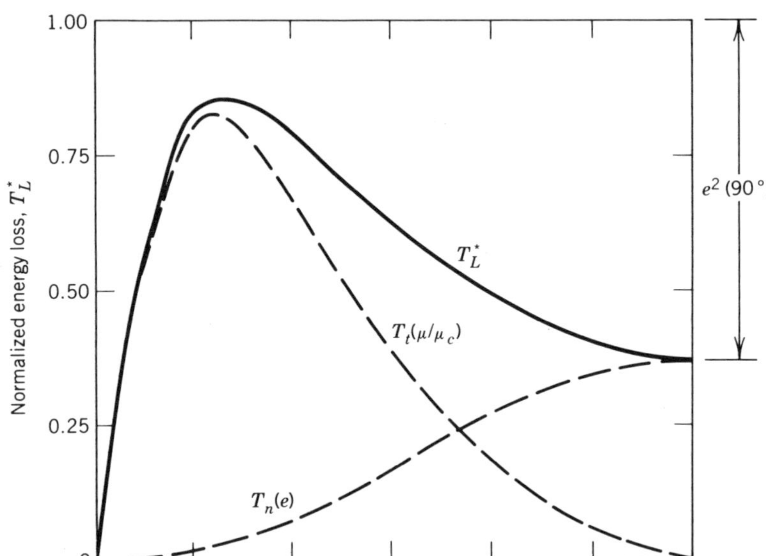


**Figure 6.9.** Variation of kinetic energy loss typically found for cutting-type erosion.

to as deformation wear. When the particles are pointed or sharp, the nature of the wear differs and is called cutting wear. These can often be distinguished by how the energy loss changes with the angle of incidence of the particles relative to thc surface. Figures 6.9 and 6.10 show typical kinetic energy losses for cutting wear and deformation wear, respectively. These curves are both plots of Eq. 6.17. The differences occur because $e$ and $\mu$ vary with $\alpha$ differently for deformation wear and cutting wear. The tangential losses associated with cutting wear are typically much higher and are predominant at lower angles of incidence. More rounded particles are involved with relatively large cratering at high angles of incidence and show correspondingly higher energy losses at these angles along with lower values of the coefficient of restitution.

An important observation to be made here is that a Coulomb friction model is not appropriate for these collisions. The particles penetrate the target material and generate a stress normal to the interfacing surface. The tangential component of this stress (or pressure) forms, or at least

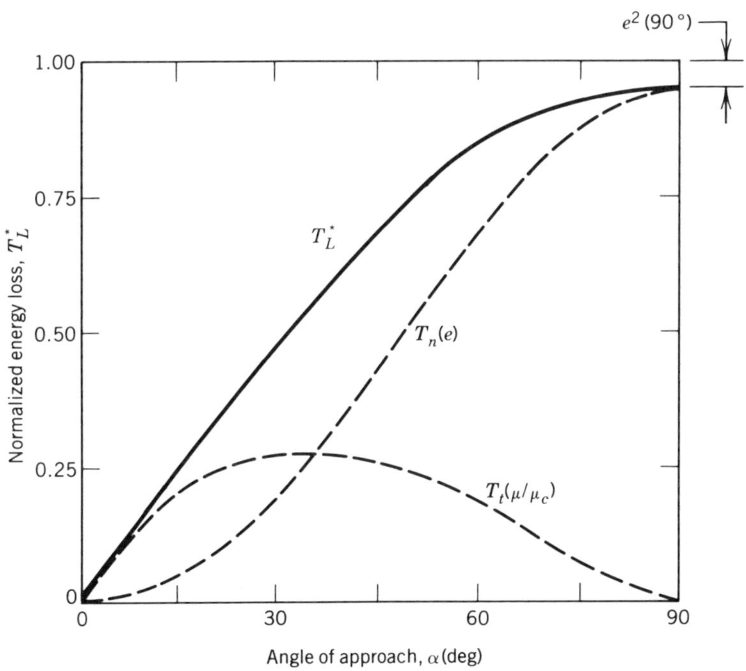


**Figure 6.10.** Variation of kinetic energy loss typically found for deformation-type erosion.

significantly contributes to, the tangential impulse. Friction here acts over the deformed, changing contact surface. In most cases plastic deformation and material removal is taking place and contributes to the tangential force component.

Figures 6.11 and 6.12 show experimental results of Neilson and Gilchrist[26] fitted with semiempirical curves containing experimental wear parameters. Note that the ordinate in these figures is erosion measured in grams of material removed (from the impacted surface) divided by the grams of solid particles propelled at the surface. Comparison of Figs. 6.11 and 6.12 with Figs. 6.9 and 6.10 shows that the process of wear due to erosion is directly tied to the kinetic energy loss of the impacting particles and that the two-parameter ($e$, $\mu$) equation, Eq. 6.17, accurately reflects the process. The relationship between energy loss and wear is well known.[10, 11]

An important characteristic found from experiments involving erosion and wear experiments is the relationship between initial energy and crater volume. Many studies show that permanent crater volume following

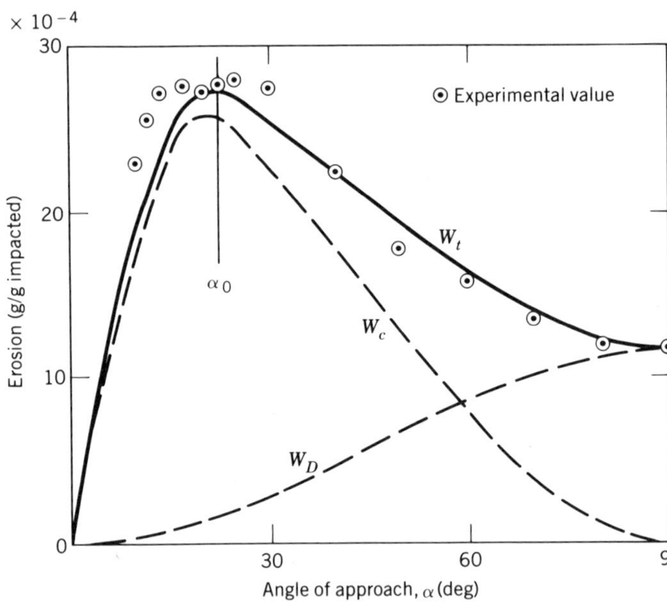


**Figure 6.11.** Normalized erosion versus angle of incidence for aluminum eroded by 210 $\mu$m aluminum oxide particles at 494 ft/s (from Ref. [26], with permission of Elsevier Sequoia S. A., copyright 1968).

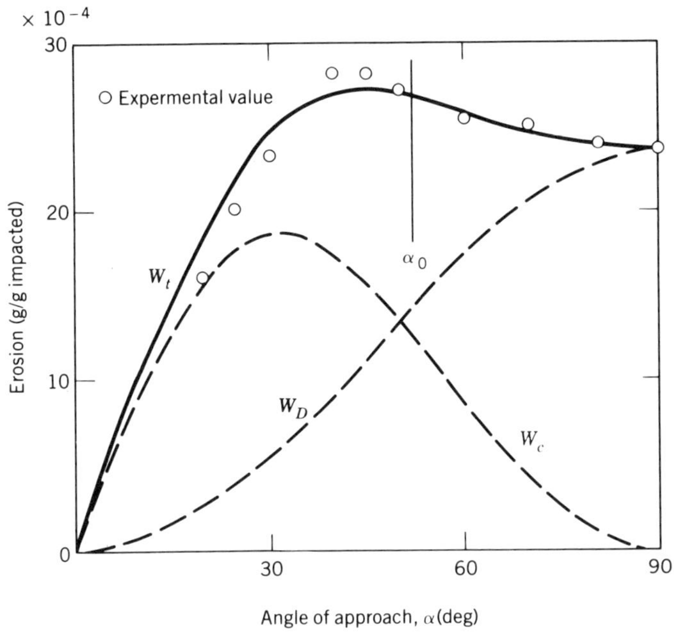


**Figure 6.12.** Normalized erosion versus angle of incidence for Perspex eroded by 210 $\mu$m aluminum oxide particles at 420 ft/s (from Ref. [26], with permission of Elsevier Sequoia S. A., copyright 1968).

impact is proportional to the velocity raised to an exponent $n$. This exponent is typically near to, but in excess of, the number 2. Since kinetic energy is proportional to the square of the velocity, this means that the volume of indentation is nearly proportional to the initial kinetic energy. Models of the local stresses and deformations have been and continue to be developed. See Refs. 9–12. Although the methods covered in this book do not model local material behavior, it is seen that the dynamics of the particles can be used to characterize the erosion process.

Some common and informative trends of the particle dynamics are observable from the data of Hutchings,[11] Sundararajan and Shewmon,[13] and others. The following observations and trends are based on data obtained primarily from oblique collisions of hardened steel spheres with massive but relatively softer metal targets and corresponds to the deformation wear category. Significant denting occurred, usually with permanent cavity formation and sometimes with removal of surface material. The velocity range of the spheres is from 0 to near 400 m/s. In summary:

1. Kinetic energy loss due to tangential effects and associated with the impulse ratio $\mu$ are most significant for angles of incidence between about 15° and 60°. Losses associated with the coefficient of restitution and due to normal effects increase monotonically with the angle of incidence from zero to the maximum values.
2. Rigid body effects (rotational inertia and energy) are not significant.
3. The tangential impulse ratio data show considerable scatter, but seem to indicate that $\mu$ is fairly constant for a given range of initial velocity.

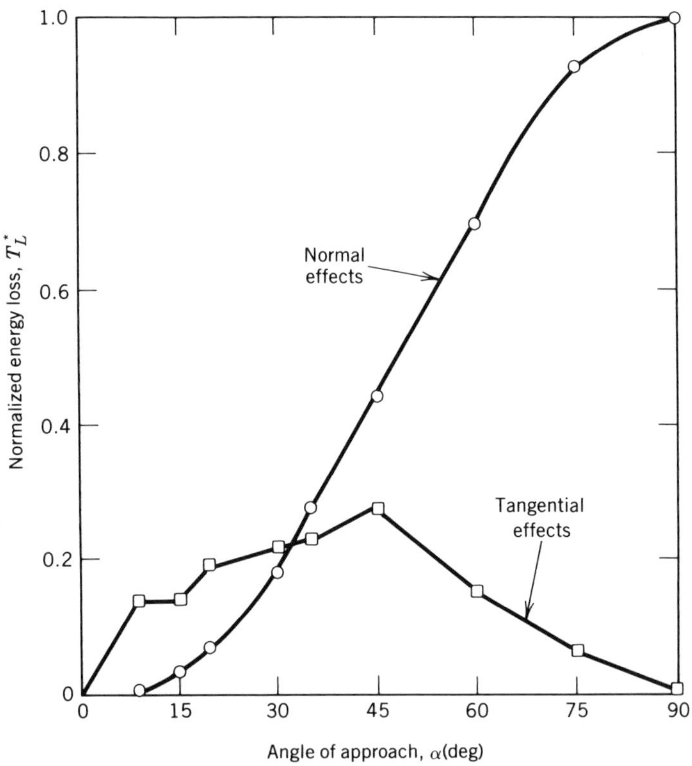


**Figure 6.13.** Experimentally measured energy loss, hard steel spheres on a mild steel target; data of Hutchings, Macmillan, and Rickerby.[12]

4. The coefficient of restitution depends jointly on the initial normal velocity and tangential velocity. It decreases rapidly with increasing initial normal velocity. However, it remains nearly constant for a given angle of incidence.

These are illustrated and discussed in more detail in the following paragraphs.

Figures 6.13–6.15 show energy loss plotted as a function of angle of incidence $\alpha$. At the lower end, for $\alpha = 0$, the normal impulse is zero and no energy can be lost. At the other extreme, $\alpha = 90°$, the impulse ratio $\mu$ must be zero and all lost energy must be related to normal deformation. In between, the tangential energy losses rise and then drop as $\alpha$ increases. Note that the stainless steels, which are harder, have considerably lower tangential energy losses compared to softer copper and brass. This is due to the shallower indentation. Mild steel lies roughly in between.

Figure 6.16 shows the impulse ratios from Hutchings[11] plotted against angle of incidence and segregated according to ranges of initial velocity. The critical value curves of impluse ratio from Eq. 6.18 are also plotted for both a sphere and a point mass ($\lambda = 0$). The data points for $\alpha =$

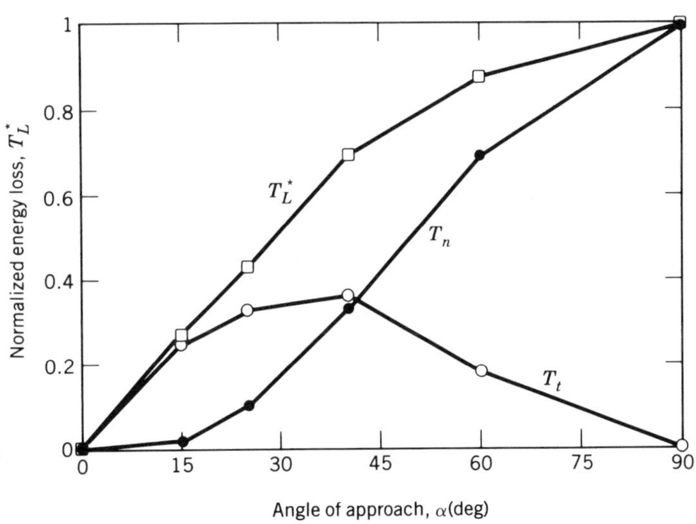


**Figure 6.14.** Experimentally measured energy loss for copper and brass; data of Sundararajan and Shewmon.[13]

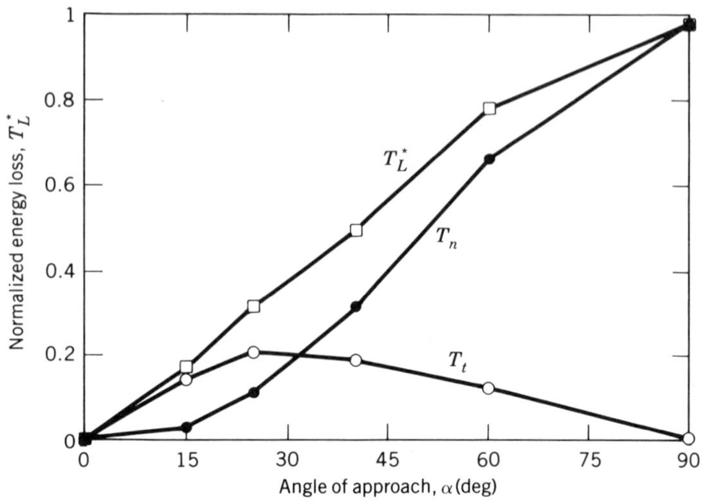


**Figure 6.15.** Experimentally measured energy loss for 301 and 304 stainless steel; data of Sundararajan and Shewmon.[13]

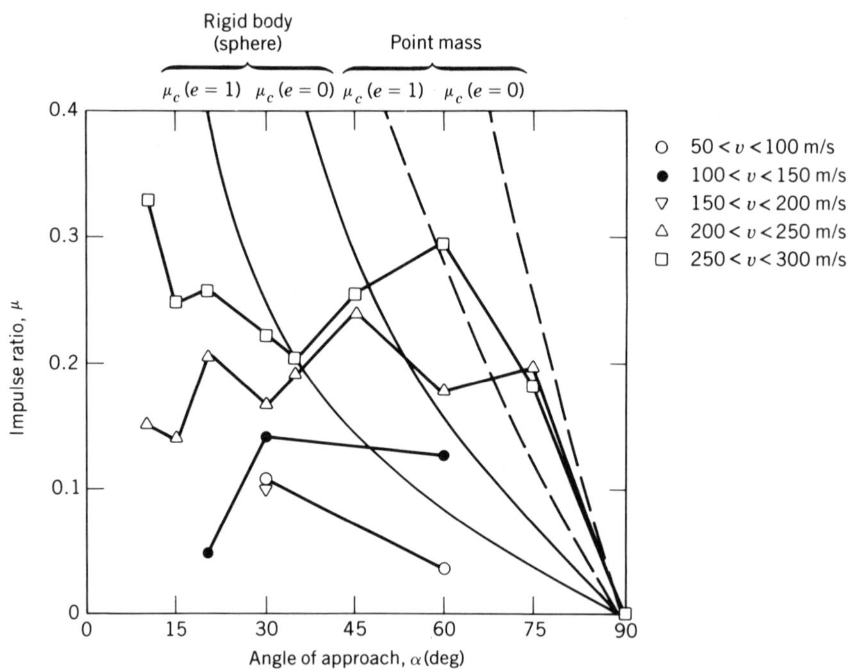


**Figure 6.16.** Experimentally measured impulse ratios, hard steel spheres on a mild steel target; data of Hutchings, Macmillan, and Rickerby.[12]

75° fall into the point-mass critical region. This corroborates the observations of Hutchings, who indicated that the rotational energy of the spheres (both before and after impact) was negligible. Consequently, the spherical projectiles behaved as point masses in these erosion and wear experiments.

Consideration of the data points of Fig. 6.16 corresponding to each individual range of initial velocity (for angles of incidence below the critical range) shows a rather wide variation. The process of crater formation typically involves material buildup at the leading edge of the crater. This can lead to the particle hopping out of the crater and even chipping of the built-up lip. For these reasons, the variations in the impulse ratio with $\alpha$ are possibly due to erosion being an erratic process rather than experimental error or measurement error. Regardless of this, there is a regular progression or trend of decreasing $\mu$ with decreasing initial velocity. This makes sense since it has already been observed that crater volume (and depth) increases with initial velocity. As the crater depth decreases, it would seem logical that $\mu$ would approach $f$ for these materials. Independent measurements[11] of $f$ yielded values in the range of 0.04–0.05. This agrees with several values of $\mu$ from individual data points, which are less than 0.1.

Data from similar experiments with materials other than mild steel[13] are plotted in Fig. 6.17. Here, the values of $\mu$ are not segregated by initial velocity, but are averaged over initial velocities ranging from about 50 to 180 m/s. Again, the data indicate point-mass behavior. But more important, values of $\mu$ below $\mu_c$ are much more uniform. Furthermore, a very high correlation exists between their average and the dynamic material hardness (experimental values are listed in parentheses in Fig. 6.17). Again, this makes sense, since the harder the material, the shallower the crater and the closer the values of $\mu$ to $f$. It is interesting that the values of $f$ observed for these materials correspond to friction coefficients typically measured under lubricated conditions.[16] This points out that it is frequently necessary to make measurements for specific applications. The above discussion indicates how the data should be analyzed and some typical trends that data tend to follow. The discussion now turns to the coefficient of restitution.

In the formulation of collision dynamics, the coefficient of restitution is defined as a kinematic constant that determines the ratio of the normal rebound velocity to the approach velocity. Its more important role is in representing the elastic/plastic behavior due to deformation in the con-

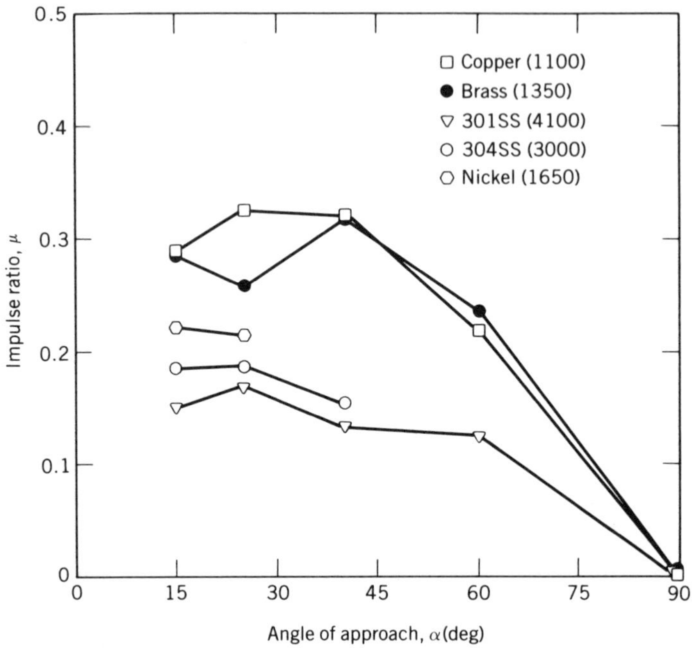


**Figure 6.17.** Experimentally measured impulse ratios; data of Sundararajan and Shewmon.[13] Numbers in parentheses are dynamic hardnesses in MPa.

tact region. As such, it controls energy loss as indicated by Eq. 6.17. As the coefficient of friction, it is a constant used to represent rather complex mechanical behavior. The accuracy of impact theory to model collisions obviously depends directly on the accuracy and behavior of this coefficient. Although treated as a constant for purposes of solving the system equations, it depends on the materials of the bodies and the magnitude of the initial velocities. It depends not only on the initial normal velocity component, but also the tangential component as well (or alternatively, on the angle of incidence). Figure 6.18 shows experimental values of $e$ for low-speed collisions of porcelain and brass spheres.[14] Typical values and trends for these and other materials are presented in other references.[1,17,18] Figure 6.19 shows how the values of $e$ vary with initial normal impact velocity, segregated according to ranges of initial tangential velocity for a hard steel sphere colliding with a massive mild steel target.[11,12] As discussed earlier, most of these collisions were accompanied by significant permanent cratering with the crater volume nearly proportional to the initial energy of the sphere. The data in both Figs. 6.18 and 6.19 can be represented approximately by a

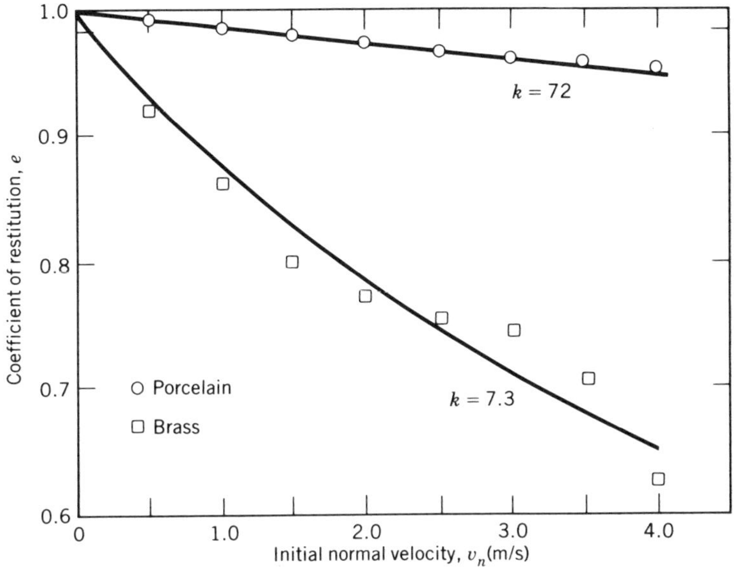


**Figure 6.18.** Coefficients of restitution; the solid line is from Eq. 6.19 (data from Tatara and Moriwaki[14] with permission of JSME).

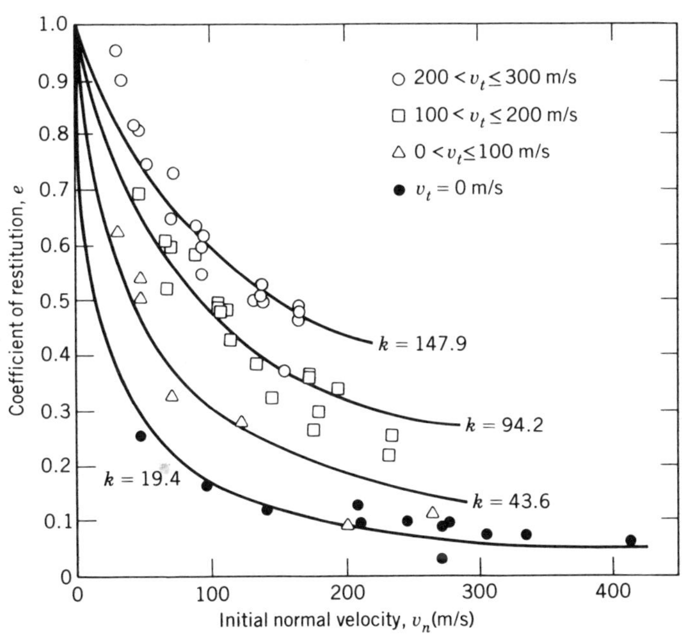


**Figure 6.19.** Coefficients of restitution for a mild steel target; data of Hutchings, Macmillan, and Rickerby.[12] The solid line is from Eq. 6.19.

relationship of the form

$$e = \frac{k}{v_n + k} \tag{6.19}$$

Experimental values of $k$ are indicated in the figures. From Fig. 6.19 it may be noted that $k$ increases directly with the initial tangential velocity. Assume for a moment that this relationship is linear, $k = bv_t$, where $b$ is a constant. Noting that $v_n/v_t = \tan\alpha$, Eq. 6.19 can be written as

$$e = \frac{b}{b + \tan\alpha} = \frac{1}{1 + a\tan\alpha} \tag{6.20}$$

where $a = 1/b$ and $\alpha$ is the angle of incidence of the spherical particle. If these assumptions are true, then the coefficient of restitution should show an explicit independence of the initial normal velocity as long as the angle of incidence is held fixed. Figures 6.20–6.22 illustrate the corresponding data of Hutchings[11] and Sundararajan and Shewmon[13] and appear to confirm the validity of Eqs. 6.19 and 6.20.

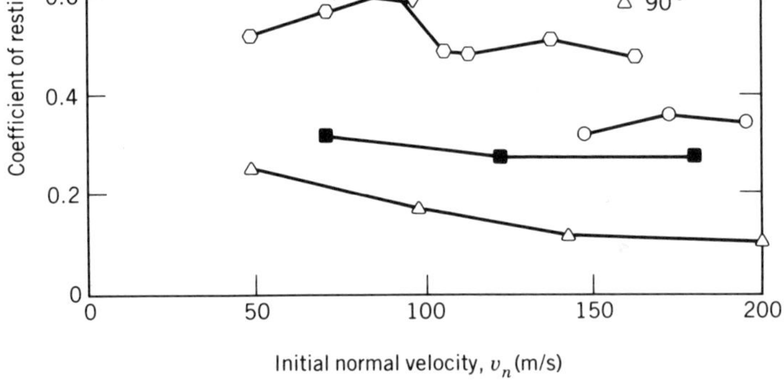


**Figure 6.20.** Coefficients of restitution for a mild steel target; data of Hutchings, Macmillan, and Rickerby.[12]

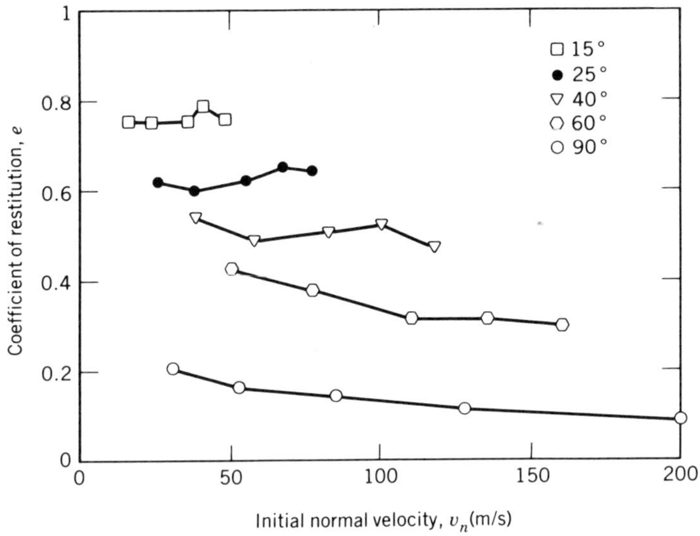


**Figure 6.21.** Coefficients of restitution for 301 and 304 stainless steels; data of Sundararajan and Shewmon.[13]

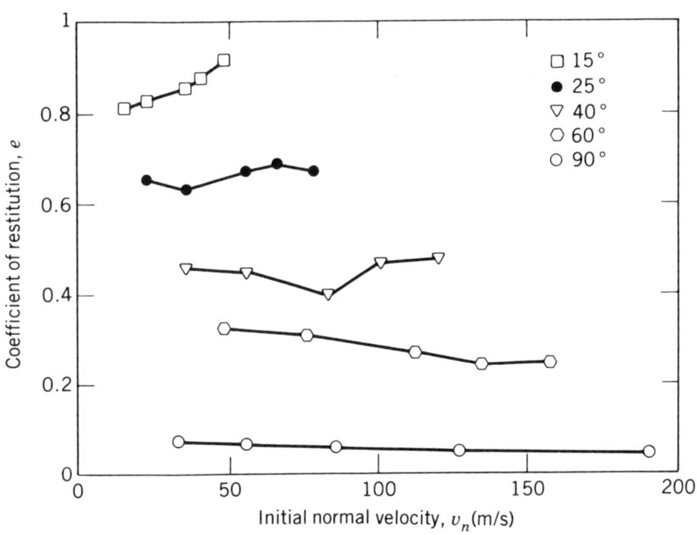


**Figure 6.22.** Coefficients of restitution for copper and brass; data of Sundararajan and Shewmon.[13]

Equation 6.20 is an empirical relationship, at this point without any theoretical foundation. Furthermore, it is approximate, since for normal impacts ($\alpha = 90°$), Eq. 6.20 predicts totally inelastic impacts, that is, $e = 0$ and does not correspond to the data. Nevertheless, it does serve to illustrate an interesting characteristic of the coefficient of restitution for this type of collision and for these materials for a relatively wide range of initial velocities. Relationships differing from Eq. 6.19 have been proposed[19] with equal validity, but this topic is not pursued any further here.

## 6.4 TANGENTIAL RESTITUTION

The concept of a tangential coefficient of restitution, $e_t$, was discussed in Chapter 2 for particle impacts. From a mathematical point of view it was found that this coefficient is well defined, solution equations can be found and that an equivalence exists between this coefficient and the impulse ratio (see Eq. 2.31). An obvious question arises as to whether this coefficient has any physical meaning or significance or both. It is conceivable that in some collisions, tangential elastic effects due to shear deformation, for example, may exist. Some experimental evidence exists[20] which indicates that tangential elastic effects do occur. These effects have been studied analytically using Hertzian theory combined with shearing loads and stresses.[21] The results of these studies have been used to explain an unusual behavior of a spinning, bouncing superball.[22] In what follows, it is shown that the concepts and equations presented thus far can be used to model adequately a collision in which elastic tangential effects exist.

The system equations developed earlier contained the definition of the coefficient $e_t$; see Eq. 6.7.3. The system equations appropriate to this problem are Eqs. 6.6, 6.7.3, and 6.8. For simplicity, it is assumed that the impulse moment $M$ is negligible, making $e_m = -1$. It is also assumed that the rigid body retains its original shape. Finally, since the experimental data[20] is for a segment of a sphere, its moment of inertia can be written as $I = mk^2$, $B = 0$, and $\lambda = C = R^2/k^2$. Under these conditions, the solution equations become

$$V_n = -ev_n \tag{6.21}$$

$$V_t = v_t - \frac{1}{1 + \lambda}(1 + e_t)(v_t - R\omega) \tag{6.22}$$

and

$$\Omega = \omega + \frac{1}{R}\frac{1}{1+\lambda}(1+e_t)(v_t - R\omega) \qquad (6.23)$$

The velocity of the point on the rigid body that initially contacts the surface is $v_t - R\omega$; the corresponding final velocity is $V_{ct}$ and is given by

$$V_{ct}(e_t) = V_t - R\Omega = -e_t(v_t - R\omega) \qquad (6.24)$$

which follows directly from the definition of $e_t$. The corresponding solution equation containing the impulse ratio is

$$V_{ct}(\mu) = V_t - R\Omega = v_t - R\omega - \mu(1+e)(1+\lambda)v_n \qquad (6.25)$$

Both Eqs. 6.24 and 6.25 are needed to model a collision with tangential restitution as will be seen shortly. First of all, however, note that an angle of incidence $\alpha$ can be defined by the equation

$$\tan\alpha = v_n/(v_t - R\omega) \qquad (6.26)$$

Based on this definition, the normalized energy loss with tangential restitution can be shown to be

$$T_L^* = (1 - e^2)\sin^2\alpha + \frac{1}{1+\lambda}(1 - e_t^2)\cos^2\alpha \qquad (6.27)$$

This is comparable to Eq. 6.17 which is based on $\mu$ and normalized to the initial energy.

Figure 6.23 shows the results of experimental measurements of Maw et al.[20,21] The data correspond to a sequence of controlled impacts of the edge of a disk (cut from a 4 in. hard steel sphere) against a flat block of like material. The coordinates of the figure are the final and initial tangential velocities of the contact point, $V_{ct}$ and $v_{ct}$, respectively, each divided by the initial normal velocity $v_n$. The theoretical curves corresponding to Eqs. 6.24 and 6.25 are laid over the data (least square fits). Equation 6.24 divided by $v_n$ corresponds to a linear relationship whose

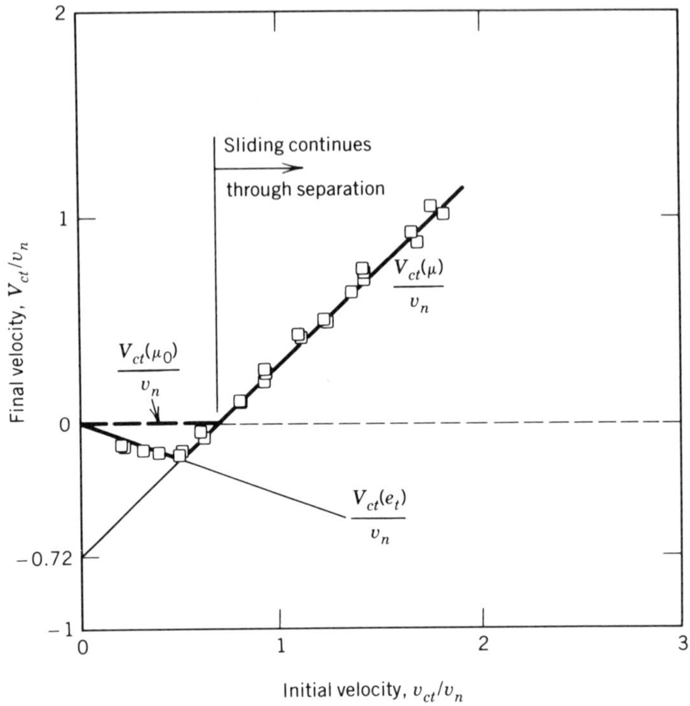


**Figure 6.23.** Final tangential velocity with tangential restitution (data of Maw, Barber, and Fawcett,[20] used with permission of Pergamon Press).

negative slope is the tangential coefficient $e_t$. Equation 6.25 divided by $v_n$ also represents a linear relationship whose slope is unity and whose intercept is $\mu(1 + e)(1 + \lambda)$. Each of these expressions fits a portion of the data fairly well to form a bilinear model. The existence of the data with negative ordinate values demonstrates the existence of tangential restitution. Without restitution, the model containing $\mu$ (Eq. 6.25) corresponds to sliding throughout the impact for values of the abscissa greater than the intercept and $V_{ct} = 0$ otherwise (dashed line). From their experiments, Maw et al. calculated an average value of $e$ to be 0.93. With $\lambda = 2$ and using the intercept of the line corresponding to $V_{ct}(\mu)$, the value of $\mu$ is determined to be 0.124. Furthermore, the slope of the $V_{ct}(e_t)$ line gives $e_t = 0.338$. Since little indentation was noted and since $\mu$ is determined from the data points with sliding throughout contact, the value of $\mu = 0.124$ corresponds to the dynamic friction coefficient $f$.

Based on the experimental work of Maw et al., it seems that, when tangential restitution occurs, it does so only for those collisions in which sliding ends prior to separation. More likely, it always occurs to some extent but is often unnoticed or insignificant or both. An almost identical phenomenon is associated with torsional motion and deformation, as is seen in the next section.

## 6.5 TORSIONAL RESTITUTION

Consider a ball spinning about a vertical axis, with velocity $v_n$ along that axis, which collides with a massive surface. Horak[23] noted that under certain conditions the spin of the ball reverses. Horak conducted experiments and developed a mathematical model of impact using the concept of a torsional coefficient of restitution. Although his model fits the data quite well, it has some deficiencies; it is independent of the moment of inertia of the spin axis and is independent of the normal impulse. Andrews,[24] without knowledge of Horak's work, uses the effects of torsional restitution to explain a certain type of handball bounce. Andrews presents an analysis that follows an assumption of no slip.

In the following, the approach of the tangential restitution analysis is followed to develop a similar bilinear model. It is compared to data from one of Horak's experiments to show that it indeed follows the trends of the data.

First consider the equations corresponding to simultaneous normal and torsional impact following a vertical fall as described above. The normal impact is characterized by the coefficient of restitution $e$, as

$$V_n = -ev_n \tag{6.28}$$

The same form of relationship can be used to represent torsional motion, that is,

$$\Omega_n = -e_z\omega_n \tag{6.29}$$

where $e_z$ is a torsional coefficient of restitution and the angular velocities are about the normal axis. (The physical origins of this coefficient and the phenomenon that it represents are not pursued here. It appears quite apparent, however, that this arises due to torsional elasticity effects, where the torsional forces are transferred through friction.) Note that Eqs. 6.28 and 6.29 are a set of solution equations. The normal impulse $P_n$ is found from the linear momentum change, Eq. 6.1. The torsional impulse is found from the change in angular momentum, that is,

$$M_n = I(\Omega_n - \omega_n) \tag{6.30}$$

Now consider the case where rotational sliding (about the normal axis) continues throughout the duration of contact. Horak's data seem to indicate that the restitutional effects continue even when sliding predominates. Thus,

$$M_n = \mu_z R_p P_n - e_z I \omega_n \tag{6.31}$$

Since sliding exists at separation, $\pm f$ is used as the impulse ratio. The quantity $R_p$ is a constant with dimensions of length and represents an equivalent moment arm or pitch radius (see Chapter 7) over which the normal impulse is transformed into the torsional impulse. Using $m(V_n - v_n)$ for $P_n$ and eliminating $M_n$ from Eqs. 6.30 and 6.31, the final angular velocity is obtained as

$$\Omega_n = (1 - e_z)\omega_n - \frac{\mu_z R_p m}{I}(1 + e)v_n \tag{6.32}$$

For sliding throughout the collision in the presence of Coulomb friction, the solution equations are Eqs. 6.28 and 6.32 with $\mu_z = \pm f$.

Expressions for energy loss are easy to obtain. For angular velocity reversal, that is, torsional restitution, the energy loss is

$$T_L = \tfrac{1}{2} m v_n^2 (1 - e^2) + \tfrac{1}{2} I \omega_n^2 (1 - e_z^2) \tag{6.33}$$

For sliding solution,

$$\begin{aligned} T_L = {} & \tfrac{1}{2} m v_n^2 (1 - e^2) + \tfrac{1}{2} I \omega_n^2 (1 - e_z^2) \\ & + \tfrac{1}{2} m v_n^2 \left\{ \frac{I}{m} \frac{\omega_n^2}{v_n^2} + 2(1 - e_z) \left[ \mu_z (1 + e) \frac{R_p \omega_n}{v_n} - \frac{I}{m} \frac{\omega_n^2}{v_n^2} \right] \right. \\ & \left. - \mu_z^2 \frac{m R_p^2}{I} (1 + e)^2 \right\} \end{aligned} \tag{6.34}$$

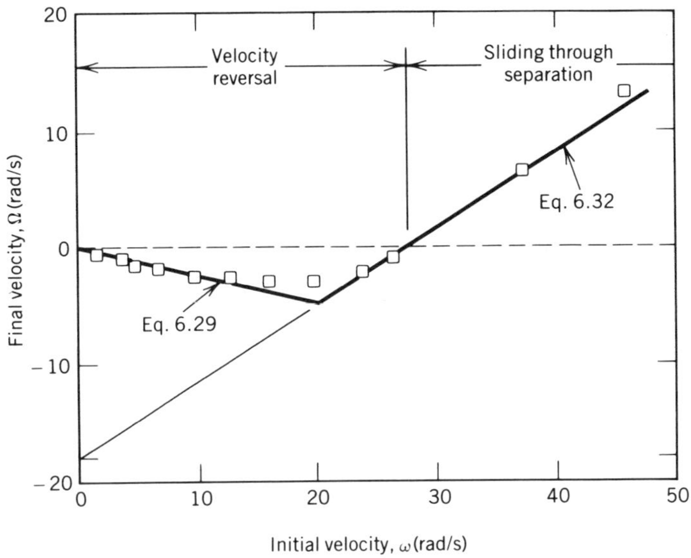


**Figure 6.24.** Final spin angular velocity with torsional restitution, with data of Horak.[23]

Figure 6.24 shows the two solution equations, Eqs. 6.29 and 6.32, laid over the experimental data of Horak, using a visual fit. The data are for a spinning rubber ball dropped vertically onto a smooth marble slab. As in the case of tangential restitution, the velocity reversal occurs over the range of initial conditions where sliding ends prior to separation. The value of $e_z$ for the experimental conditions is approximately 0.24. The sliding solution's intercept is used to obtain a value of $R_p = 0.21$ in. This follows from measured values of $e = 0.82$ and $f = 0.24$.[23] For values of $\omega_n = 27.5$ rad/s sliding continues through separation.

## 6.6 THE MOMENT COEFFICIENT

The moment coefficient $e_m$ was introduced in Chapter 5 (see Eq. 5.11). This coefficient was introduced because of the potential for the physical presence of a moment and its resulting impulse over the contact surfaces. One of the properties of $e_m$ was illustrated in Example 5.1, where $e_m = 0$ was seen to cause the rigid bodies to have a common rotational velocity following impact. The same example shows that $e_m = -1$ provides a solution with no moment impulse, $M = 0$. These observations relate to the kinematical properties of $e_m$. In this section, the moment coefficient and its relationship to elastic properties are explored.

### 6.6.1 An Offset Normal Impulse

Before reviewing impact equations, consider first some concepts from elementary mechanics concerning equivalent-force systems. Figure 6.25 shows a one-dimensional distributed contact force $w(x)$ between a rigid block and a flexible surface. The force $w(x)$ can be replaced by an equivalent point force $F$ at a distance $x$ and a moment $N$ provided that two conditions are met. These conditions are:

1. that $F$ is equal to the resultant force of $w(x)$, and
2. that the combined moment of $F$ and $N$ about an arbitrary point is equal to the corresponding moment of $w(x)$.

Furthermore, if the product $Fd$ is equal to the moment of $w(x)$ and if the line of action of $F$ is at $x = d$, then $N = 0$. The same conclusions apply to impulses $P_n$ and $M$ as to $F$ and $N$, respectively. If an equivalent impulse system consisting of $P_n$ and $M$ acts at point $C$ and point $C$ is

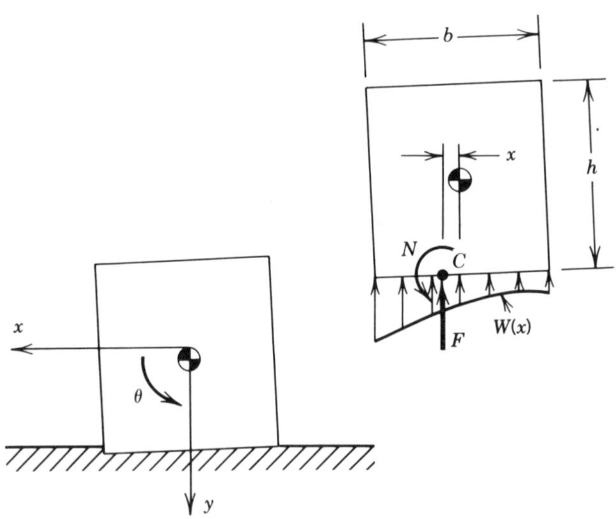


**Figure 6.25.** Diagrams of falling block problem.

chosen such that the impulse of a distributed surface force $w(x, t)$ has no impulse moment about $C$, then $M = 0$. The use of any other point requires that $M$ be nonzero.

When analyzing collisions that have a considerable area of contact, the specific location of the line of action of $P_n$ for which $M = 0$, often is unknown. Either an assumption that $M = 0$ must be made or an unknown moment impulse $M$ must be included in the problem. Consequently, an impulse moment $M$ can exist in a collision analysis because of an inappropriate choice of the line of action of $P_n$ or it can exist because the line of action of $P_n$ is known and an impulse moment is physically present due to the force system. Unfortunately, the equivalence of force systems (or impulse systems) does not ensure the equivalence or identity of responses. To see this, consider the problem of a dropped block with an offset normal impulse.

Suppose a uniform rectangular solid falls onto a resilient surface containing an imperfection, such as a void beneath the surface. The imperfection has the effect of causing an offset $d$ to the normal impulse as shown in Fig. 6.26$a$. It is not difficult to show that the solution equations are

$$V_n = v_n - q(1 + e)(v_n + d\omega) \tag{6.35}$$

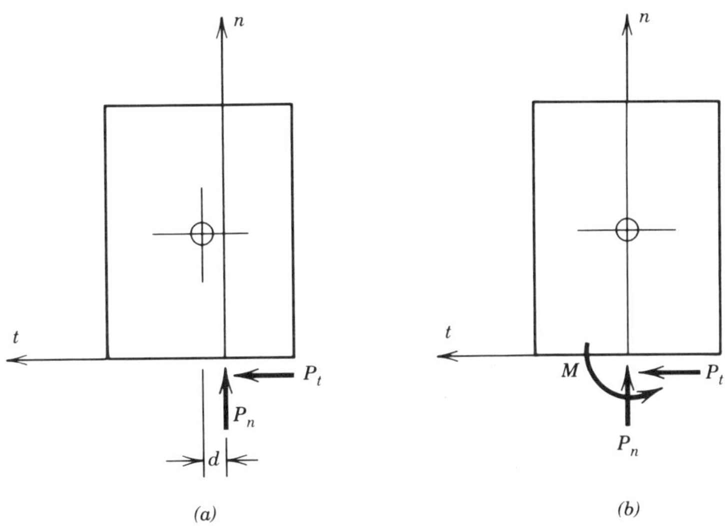


**Figure 6.26.** Equivalent rigid body impulse systems at interface of falling block.

$$V_t = v_t - \mu q(1 + e)(v_n + d\omega) \tag{6.36}$$

and

$$\Omega = \omega - q(1 + e)\frac{d - \mu h/2}{k^2}(v_n + d\omega) \tag{6.37}$$

where

$$q = \left(1 + \frac{d(d - \mu h/2)}{k^2}\right)^{-1} \tag{6.38}$$

Now consider the possibility that existence of $d$ is unknown to an analyst and the central impact problem of Fig. 6.26*b* is used in the analysis. Furthermore, the solution is not sought using the moment coefficient $e_m$. From the discussions of tangential and torsional restitution, it has been learned that if "restitution," that is, an elastic effect, does not exist, an impulse ratio is more appropriate. For this reason, instead of $e_m$, an impulse ratio is defined, where

$$\lambda_M = M/P_n \tag{6.39}$$

is used as one of the system equations in place of Eq. 6.8. The solution equations are

$$V_n = v_n - q_M(1 + e)(v_n + d_c\omega) \tag{6.40}$$

$$V_t = v_t - \mu q_M(1 + e)(v_n + d_c\omega) \tag{6.41}$$

and

$$\Omega = \omega - q_M(1 + e)\frac{\lambda_M + d_c - \mu d_c}{k^2}(v_n + d_c\omega) \tag{6.42}$$

where

$$q_M = \left(1 + \frac{\lambda_M d_c + d_c^2 - \mu d_c d_d}{k^2}\right)^{-1} \tag{6.43}$$

For this latter solution to be equivalent to the former, its impulse moment $M$ must equal the impulse $P_n$ multiplied by the offset $d$. That is,

$$\lambda_M q_M(1 + e)(v_n + d_c\omega)\big|_{d_c=0} = dq(1 + e)(v_n + d\omega) \quad (6.44)$$

But $q_M(d_c = 0) = 1$, so that

$$\lambda_M = dq\,\frac{v_n + d\omega}{v_n} \quad (6.45)$$

where $q$ is given by Eq. 6.38. Although a value of $\lambda_M$ has been determined for an equivalent impulse system, the solutions are not the same. Comparison of $V_n$ from Eq. 6.35 and $V_n$ from Eq. 6.40 shows different final normal velocities. This difference, in turn, implies different normal impulses. So, although the moments are equivalent, the solutions are not.

In some cases, the solutions can be the same, however. Consider a vertical drop with $v_t = \omega = 0$ and a small offset $d = \epsilon \approx 0$. Under these conditions $|\mu| \leq |\mu_0| = 0$. Furthermore, since $\epsilon^2 << k^2$, $\lambda_M \approx \epsilon$, and to a first-order approximation, both solutions are identical.

### 6.6.2 Dropped Block on a Surface With Linearly Varying Mechanical Properties

Some approximate relationships between the coefficient of restitution $e$ and the elastic and dissipative material properties of the colliding bodies were explored in Chapter 3. When rotational restitution occurs, it too should be relatable to the local material properties. An example is presented here of a rigid rectangular block falling onto a surface whose mechanical properties vary linearly from side to side of the block.[25]

Suppose the flat surface has elastic and dissipative properties defined by the following equations:

$$k(x) = k_0 + k_x x \quad (6.46)$$

and

$$c(x) = c_0 + c_x x \quad (6.47)$$

The quantity $k(x)$ is a distributed surface stiffness and $c(x)$ a distributed surface damping coefficient. An inertial $x - y$ coordinate system is shown in Fig. 6.25, where $y$ and $x$ indicate mass center positions of the block and $\theta$ is its angular rotation measured positive counterclockwise. Methods typically used in linear vibration theory allow the response of the block to be calculated. Several assumptions are made to simplify the problem; these are

1. displacements $y(\tau)$ and $\theta(\tau)$ are small and $x(\tau)$ is negligible;
2. damping is below critical;
3. the natural period of rotational motion is greater than that of vertical motion; and
4. the duration of contact is one-half of the natural period of vertical motion.

For small displacements the differential equations of motion can be linearized. This results in a two-degree-of-freedom system with equations

$$m\ddot{y} + bc_0\dot{y} - \frac{b^3}{12}c_x\dot{\theta} + bk_0 y - \frac{b^3}{12}k_x\theta = 0 \qquad (6.48)$$

and

$$12\frac{I}{b^3}\ddot{\theta} - c_x\dot{y} + c_0\dot{\theta} - k_x y + k_0\theta = 0 \qquad (6.49)$$

where the overdots represent derivatives with respect to time $\tau$. Let $\omega_y$ and $\omega_\theta$ be the eigenvalues or natural frequencies. The duration of contact is $\pi/\omega_y$. The impulse moment can be calculated. This is

$$\begin{aligned} M &= \frac{b^3}{12}\int_0^{\pi/\omega_y}(k_0\theta - k_x y)\,d\tau + \frac{b^3}{12}\int_0^{\pi/\omega_y}(c_0\dot{\theta} - c_x\dot{y})\,d\tau \\ &= \frac{k_x b^3 v_0(b^2 + h^2)}{12\omega_y^2 h^2}\left(1 + \cos\beta - \frac{c_0\omega_\theta}{k_0}\sin\beta\right) \end{aligned} \qquad (6.50)$$

where $\beta = \pi\omega_\theta/\omega_y$. For a vertical drop, $\dot{x}_0 = v_t = 0$ and $\dot{\theta}_0 = \omega = 0$. Equation 6.8 can be solved for $e_m$, the moment coefficient. This is

$$e_m = -\frac{I\Omega}{I\Omega - M} \tag{6.51}$$

But $\Omega = \dot{\theta}(\tau/\omega_y)$, so that

$$\Omega = \frac{v_0 k_x b^2}{k_0 h^2}(1 + \cos\beta) \tag{6.52}$$

Consequently,

$$e_m = \frac{1}{\dfrac{c_0\omega_\theta}{k_0}\dfrac{\sin\beta}{1 + \cos\beta} - 2} \tag{6.53}$$

Note also that for this system, the coefficient of restitution is

$$e = \exp\left(\frac{-\pi b c_0}{2\sqrt{bk_0 m}}\right) \tag{6.54}$$

The previous discussion highlights some properties of the coefficient $e_m$. Being newly defined, however, more exploration of its properties and utility is required.

## 6.7 PROBLEMS

**6.1** Verify Eq. 6.7.

**6.2** Show that for a spherical particle, $\mu_0 < \mu_T$, where $\mu_0$ is the impulse ratio sufficient to cause sliding to end at the time of separation and $\mu_T$ is the impulse ratio (greater than zero) for zero energy loss.

**6.3** A circular cylinder of radius $R$ rolls toward and strikes a step of height $h < R$ with a velocity $v_0 = R\omega_0$. Show that the maximum impulse ratio $\mu_m$ is

$$\mu_m = \frac{e_m + \sin\beta}{(2e_m - 1)(1 + e)\cos\beta}$$

where $\sin\beta = (R - h)/R$

**6.4** Two coaxially rotating rigid bodies have moments of inertia $I_1$ and $I_2$. They are brought into contact with small initial angular velocities $\omega_1$ and $\omega_2$, respectively. Under the assumption that torsional restitution occurs, show that the final angular velocities are

$$\Omega_i = \omega_j + \bar{I}(1 + e_z)(\omega_2 - \omega_1)/I_i, \qquad i = 1, 2$$

**6.5** It is shown in Chapter 1 that for point-mass impacts the kinematic definition of the coefficient of restitution, Eq. 1.15, is equivalent to the kinetic definition, Eq. 1.14, where the latter is $e = P_R/P_A$. For a rigid body barrier impact, show that the same is true if $d_c = 0$ or if the approach and rebound impulse ratios $\mu_A$ and $\mu_R$ are equal.

**6.6** Show for an impact of any two bodies that the kinetic definition of $e$, Eq. 1.14, leads to the relationship

$$\mu = \frac{\mu_A + e\mu_R}{1 + e}$$

where $\mu_A$ and $\mu_R$ are the approach and rebound impulse ratios, respectively, and are given by $\mu_A = P_t^A/P_n^A$ and $\mu_R = P_t^R/P_n^R$, respectively.

## REFERENCES


1. Keller, J. B., "Impact with Friction," *J. Appl. Mech.*, **53,** 1, 1986.
2. Goldsmith, W., *Impact*, Edw. Arnold, London 1960.
3. Macaulay, M., *Introduction to Impact Engineering*, Chapman and Hall, London, 1987.
4. Kilmister, C. W. and Reeve, J. E., *Rational Mechanics*, Longmans, London, 1966.

5. Shames, I. H., *Engineering Mechanics, Statics*, Prentice-Hall, Englewood Cliffs, NJ, 1966.
6. Peterson, I., "Friction Features," *Science News*, **133**(18), 283, 1988.
7. Ratner, S. B. and Styller, E. E., "Characteristics of Impact Friction and Wear of Polymeric Materials," *Wear*, **73,** 213–234, 1981.
8. Rabinowicz, E., *Friction and Wear of Materials*, John Wiley, New York, 1965.
9. Engel, P. A., *Impact Wear of Materials*, Elsevier, Amsterdam, 1976.
10. Adler, W. F., *Erosion: Prevention and Useful Applications*, ASTM STP 664, ASTM, Philadelphia, PA, 1979.
11. Hutchings, I. M., "The Erosion of Ductile Metals by Solid Particles," Ph.D. Dissertation, Trinity College, Cambridge, 1974.
12. Hutchings, I. M., Macmillan, N. H., and Rickerby, D. G., "Further Studies of the Oblique Impact of a Hard Sphere Against a Ductile Solid," *Int. J. Mech. Sci.*, **23,** 639–643, 1981.
13. Sundararajan, G. and Shewmon, P. G., "The Oblique Impact of a Hard Ball Against Ductile, Semi-Infinite Target Materials—Experiment and Analysis," *Int. J. Impact Engrg.*, **6,** 3–22, 1987.
14. Tatara, Y. and Moriwaki, N., "Study on Impact of Equivalent Two Bodies," *Jpn. Soc. Mech. Engr.*, **25**(202), 631–638, 1982.
15. Brach, R. M., "Impact Dynamics with Applications to Solid Particle Erosion," *Int. J. of Impact Engrg.*, **7**(1), 37–54, 1988.
16. Baumeister, T. (ed.), *Mechanical Engineer's Handbook*, 6th ed., McGraw-Hill, New York, 1985.
17. Barkan, P., "Impact," *Mechanical Design and Systems Handbook*, Section 16, McGraw-Hill, New York, 1974.
18. Higdon, A., et al., *Engineering Mechanics*, Prentice-Hall, Englewood Cliffs, NJ, 1979.
19. Hunt, K. H. and Crossley, F. R. E., "Coefficient of Restitution Interpreted as Damping in Vibroimpact," *Trans. ASME, J. Appl. Mech.*, 440–445, 1975.
20. Maw, N., Barber, J. R., and Fawcett, J. N., "The Rebound of Elastic Bodies in Oblique Impact," *Mech. Res. Comm.*, **4,** 17–21, 1977.
21. Maw, N., Barber, J. R., and Fawcett, J. N., "The Role of Elastic Tangential Compliance in Oblique Impact," *Trans. ASME, J. Lub. Technology*, **103**(74), 74–80, 1981.
22. Johnson, K. L., "The Bounce of 'Superball,'" *Int. J. Mech. Engrg. Educ.*, **11**(1), 57–63, 1983.

23. Horak, F., ''Impact of a Rough Ball Spinning Round its Vertical Diameter onto a Horizontal Plane,'' *Trans. Faculty of Mech. and Elec. Engrg.*, Technical Univ. of Prague, Czechoslovakia, 1948.
24. Andrews, J. G., ''A Mechanical Analysis of a Special Class of Rebound Phenomena,'' *Medicine and Science in Sports and Exercise*, **15**(3), 256–266, 1983.
25. Brach, R. M., ''Moments Between Impacting Rigid Bodies,'' *J. Mech. Design, ASME Trans.*, **103,** 812–817, 1981.
26. Neilson, J. H. and Gilchrist, A., ''Erosion by a Stream of Solid Particles,'' *Wear*, **II,** 111–122, 1968.

## CHAPTER 7

# THREE-DIMENSIONAL IMPACT OF RIGID BODIES

Nothing fundamentally new about impact dynamics is introduced in this chapter. All of the basic concepts already covered are assembled here to deal with the three-dimensional problem. Each rigid body has six degrees-of-freedom and will have six velocity components following an impact. This produces 12 unknown velocity components for a collision of two bodies. If unknown impulse components are not introduced, the count stays at 12. For a point impact a common way of developing the system equations[1] is the following:

- Conservation of angular momentum of each body about orthogonal axes through the contact point; six equations.
- Conservation of linear momentum of the system of two bodies along an axis normal to the impact; one equation.
- Conservation of linear momentum of the system of two bodies along two axes tangent to the impact; two equations.
- Frictional impulses, zero velocity constraints, or conservation of linear momentum for each body (frictionless case) along two axes tangent to the impact; two equations.
- Restitution or energy conservation along an axis normal to the impact; one equation.

This process produces alternative sets of 12 equations depending on which tangential equations are chosen. However, this paradigm does not apply if the impact cannot be assumed to occur at a point and does not cover cases where restitution occurs other than normal to the impact.

A more general view of the three-dimensional problem can be taken, which includes impulses as unknowns and does away with the assumption of point impact. In addition to the 12 final velocity components, the unknowns include three force impulse components and three moment impulse components. The source of the 18 system equations is the following:

- Linear impulse and momentum equations for each body; six equations.
- Angular impulse and momentum for each body; six equations.
- Normal restitution; one equation.
- Tangential impulse ratios; two equations.
- Rotational restitution about tangential axes; two equations.
- Torsional impulse ratio (torsional friction); one equation.

If tangential restitution exists, then two restitution equations replace the two tangential impulse ratio equations (or a bilinear model can be used as discussed in Chapter 6). Torsional restitution (or a bilinear model) can also replace the equation arising from the torsional impulse ratio.

Because of the alternatives in tangential and torsional restitution or friction (impulse ratios), one general set of system equations cannot be written. Conditions corresponding to no sliding at separation also could lead to a different set of system equations. A set of system equations using impulse ratios is derived and the alternatives are discussed in this chapter. Because of the number and complexity of the system equations, computer solutions are typically necessary. One relatively simple example is presented.

## 7.1 SYSTEM EQUATIONS

Figure 7.1 shows an inertial coordinate system $(n, t, t')$ whose origin, for convenience, is chosen to be attached at an appropriate point, $C$, in the contact region of the two impacting rigid bodies. Ideally, point $C$ is

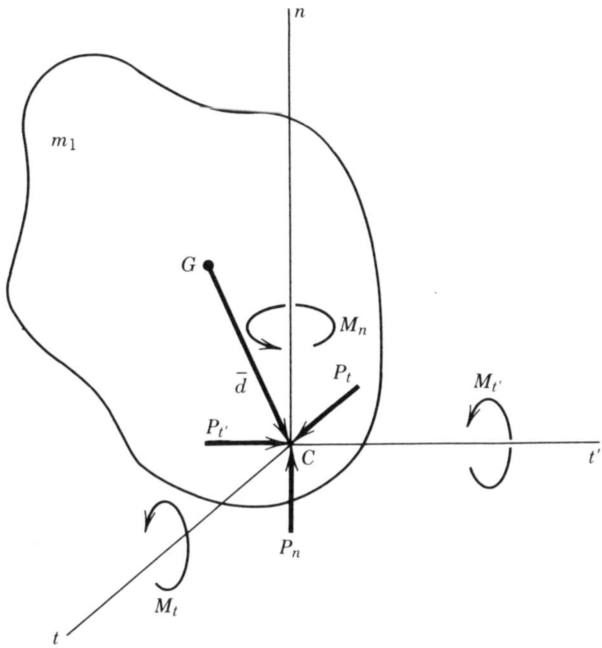


**Figure 7.1.** Coordinate system and impulse components.

located at a point on the line of action of the resultant impulse vector of the impact forces. The exact location of this point may be unknown and may need to be established by experiment or by assumption. It may or may not be on either body's undeformed or final surfaces (if they differ). Typically, the initial point of contact is chosen if deformations are small. The correct point is likely to be inside the undeformed outer surface of each body, particularly when deformations are significant. The coordinate system and impulses shown in Fig. 7.1 are assumed to be attached to a point *C*, common to both bodies with the normal axis pointing away from body 2 and toward body 1. For simplicity body 2 is not shown. Equal and opposite force and moment impulse components act between the bodies.

The two rigid bodies are assumed to be unconstrained. If either contains a kinematic constraint such as a revolute joint or slot, additional equations of constraint are needed, a topic deferred to Chapter 8. Impulses of all forces other than the contact force are omitted as are permanent shape changes in each body. When known, external impulses can be introduced.

Based on fundamental principals, the change in linear momentum is equal to the impulse on each body. This provides two vector* equations:

$$m_1(\mathbf{V}_1 - \mathbf{v}_1) = \mathbf{P} \tag{7.1}$$

and

$$m_2(\mathbf{V}_2 - \mathbf{v}_2) = -\mathbf{P} \tag{7.2}$$

where the internal impulse is given by

$$\mathbf{P} = P_n\mathbf{n} + P_t\mathbf{t} + P_{t'}\mathbf{t}' \tag{7.3}$$

and is illustrated in Fig. 7.2. The quantities $\mathbf{n}$, $\mathbf{t}$, and $\mathbf{t}'$ are unit vectors along the corresponding axes. The change in angular momentum must equal the sum of impulse moments. This gives

$$\mathbf{H}_1 - \mathbf{h}_1 = \mathbf{d}_1 \times \mathbf{P} + \mathbf{M} \tag{7.4}$$

and

$$\mathbf{H}_2 - \mathbf{h}_2 = \mathbf{d}_2 \times \mathbf{P} - \mathbf{M} \tag{7.5}$$

where $\mathbf{H}$ and $\mathbf{h}$ represent the angular momentum of a rigid body about axes through the mass center, each $\mathbf{d}$ is a position vector from a mass center to point $C$, and $\mathbf{M}$ is the internal moment impulse whose components are illustrated in Fig. 7.1.

The final and initial relative normal velocity at point $C$ are related by the coefficient of restitution. The corresponding equation can be written

$$\mathbf{V_r} \cdot \mathbf{n} = -e\mathbf{v_r} \cdot \mathbf{n} \tag{7.6}$$

Classically, $0 \leq e \leq 1$; however, as discussed in Chapter 2, negative values can be physically meaningful. The vectors

$$\mathbf{V_r} = \mathbf{V}_2 + \mathbf{\Omega}_2 \times \mathbf{d}_2 - \mathbf{V}_1 - \mathbf{\Omega}_1 \times \mathbf{d}_1 \tag{7.7}$$

*Bold face type indicates a vector quantity.

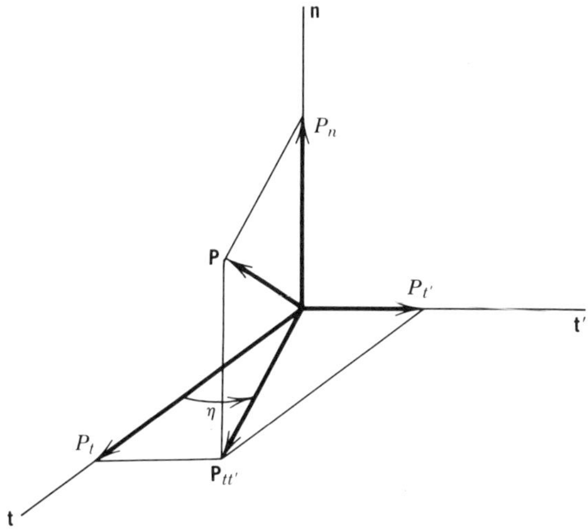


**Figure 7.2.** Force impulse components.

and

$$\mathbf{v}_r = \mathbf{v}_2 + \boldsymbol{\omega}_2 \times \mathbf{d}_2 - \mathbf{v}_1 - \boldsymbol{\omega}_1 \times \mathbf{d}_1 \tag{7.8}$$

are the relative final and initial velocities respectively, at point $C$. The tangential to normal impulse ratios, $\mu_t$ and $\mu_{t'}$, correspond to the tangential impulse components $P_t$ and $P_{t'}$, respectively, and are related by

$$\mathbf{P} = P_n(\mathbf{n} + \mu_t \mathbf{t} + \mu_{t'} \mathbf{t}') \tag{7.9}$$

In general, the impulse ratios $\mu_t$ and $\mu_{t'}$ are assumed to be independent constants. But if the tangential surface force is isotropic, then

$$\mu_t = \mu \cos \eta$$

and

$$\mu_{t'} = \mu \sin \eta \tag{7.10}$$

where $\eta$ is the angle between the $t$ axis and the projection, $\mathbf{P}_{tt'}$, of $\mathbf{P}$ onto the $t$–$t'$ plane (see Fig. 7.2). Under these conditions the tangential vector impulse is

$$\mathbf{P}_{\mathbf{tt}'} = \mu \cos \eta \mathbf{t} + \mu \sin \eta \mathbf{t}' \tag{7.11}$$

and $P_{tt'} = \mu P_n$.

The rotational effects of the interface deformation can be modeled by three independent moment coefficients. The equation that defines these coefficients can be written as

$$\begin{aligned}\left[e_{mn} M_n + (1 + e_{mn})(\Omega_{2n} - \Omega_{1n})\right]\mathbf{n} + \left[e_{mt} M_t + (1 + e_{mt})\right.\\ \left.\cdot (\Omega_{2t} - \Omega_{1t})\right]\mathbf{t} + \left[e_{mt'} M_{t'} + (1 + e_{mt'})(\Omega_{2t'} - \Omega_{1t'})\right]\mathbf{t}' = \mathbf{0}\end{aligned} \tag{7.12}$$

As in the planar case, if any of these moment coefficients is equal to $-1$, then the corresponding moment impulse is zero. Otherwise, $-1 < e_{m_i} \leq 1$.

Equations 7.1, 7.2, 7.4–7.6, 7.9, and 7.12 form a set of 18 linear scalar system equations whose solution furnishes the unknowns enumerated previously. As for many three-dimensional dynamics problems, Eqs. 7.1, 7.2, 7.4, and 7.5 are usually more conveniently expressed in terms of a body-fixed coordinate system. Consequently, a coordinate transformation may be desirable.

Other forms of system equations can be necessary, such as ones that correspond to tangential restitution rather than friction. In this case, Eq. 7.9 must be replaced by equations that define tangential restitution coefficients $e_t$ and $e_{t'}$. Two appropriate scalar equations are

$$\mathbf{V}_{\mathbf{r}} \cdot \mathbf{t} = -e_t \mathbf{v}_{\mathbf{r}} \cdot \mathbf{t} \tag{7.13}$$

and

$$\mathbf{V}_{\mathbf{r}} \cdot \mathbf{t}' = -e_{t'} \mathbf{v}_{\mathbf{r}} \cdot \mathbf{t}' \tag{7.14}$$

When relative tangential motion ends at or prior to separation, Eq. 7.9 is replaced by

$$\mathbf{V}_{ct} = V_{ct}\mathbf{t} + V_{ct'}\mathbf{t}' = 0 \tag{7.15}$$

From solutions with this condition, the ratios of $P_t$ and $P_{t'}$ to $P_n$ provide values of $\mu_{t0}$ and $\mu_{t'0}$, respectively.

A bilinear model combining restitution and friction may be appropriate, such as presented in Chapter 6. Equations 7.13 and 7.14 are used for low values of initial tangential velocity and Eq. 7.9 is used as initial tangential velocities increase and the region is approached where sliding exists at separation.

The same concepts that govern tangential motion apply to torsional or spinning motion, that is, rotation about the normal axis. Here, too, a bilinear model is the most general. However, relative rotational sliding is not commonly discussed and its treatment requires some discussion. When relative spin occurs about the normal axis, an appropriate equation is

$$M_n = -\mu_n R_p P_n \tag{7.16}$$

Because of the difference in units between a torque and force, Eq. 7.16 requires a characteristic pitch radius, $R_p$, so that the coefficient $\mu_n$ remains dimensionless. This pitch radius is assumed to be a known constant. It depends on the contact area and pressure distribution, both of which vary over time during an impact. The value of $R_p$ would typically be established experimentally, but in some simple cases it may be estimated analytically. For a contact region with an annular shape, estimates of the pitch radius may be established from clutch and brake theory.[2] Let the outside radius be $R_0$ and inside radius is $R_i$. For a uniform pressure distribution

$$R_p = \frac{2(R_o^3 - R_i^3)}{3(R_o^2 - R_i^2)} \tag{7.17}$$

For rotational sliding with a uniformly distributed power loss (uniform wear theory),

$$R_p = \tfrac{1}{2}(R_o + R_i) \tag{7.18}$$

In all cases, the pitch radius represents a characteristic distance which relates the frictional torque impulse to the normal force impulse. Expressions other than these must be established for noncircular contact regions and where the moment is due to a process of force and torque generation other than surface friction. Note that $R_p$ may be related directly to the amount of deformation which, in turn, is related to the initial relative normal velocity, that is, $R_p = R_p(v_n)$.

### 7.1.1 Constrained Motion and External Impulses

Cases often arise when one or both bodies have a motion constraint such as a pendulum with a point of fixed rotation away from the contact region. In addition, external impulses may not always be negligible. Equations 7.1–7.5 can be modified easily to take these into account. They become

$$m_1(\mathbf{V}_1 - \mathbf{v}_1) = \mathbf{P} + \mathbf{C}_1 + \mathbf{P}_1 \tag{7.19}$$

$$m_2(\mathbf{V}_2 - \mathbf{v}_2) = -\mathbf{P} + \mathbf{C}_2 + \mathbf{P}_2 \tag{7.20}$$

$$\mathbf{H}_1 - \mathbf{h}_1 = \mathbf{d}_1 \times \mathbf{P} + \mathbf{d}_{\mathbf{C}_1} \times \mathbf{C}_1 + \mathbf{d}_{\mathbf{P}_1} \times \mathbf{P}_1 + \mathbf{M} \tag{7.21}$$

and

$$\mathbf{H}_2 - \mathbf{h}_2 = \mathbf{d}_2 \times \mathbf{P} + \mathbf{d}_{\mathbf{C}_2} \times \mathbf{C}_2 + \mathbf{d}_{\mathbf{P}_2} \times \mathbf{P}_2 - \mathbf{M} \tag{7.22}$$

The quantities $\mathbf{P}_1$ and $\mathbf{P}_2$ are the external impulses on bodies 1 and 2, respectively. They are presumed to be known and act at points located by vectors $\mathbf{d}_{\mathbf{P}_1}$ and $\mathbf{d}_{\mathbf{P}_2}$ originating at each respective body's mass center. The vectors $\mathbf{C}_1$ and $\mathbf{C}_2$ are unknown impulses associated with a constraint at respective points located by centroidal vectors $\mathbf{d}_{\mathbf{C}_1}$ and $\mathbf{d}_{\mathbf{C}_2}$ on bodies 1 and 2, respectively. When a constraint exists on a rigid body, the presence of the unknown impulse requires an additional system equation. These constraint equations will ordinarily be linear functions of the velocities. For example, if body 1 has a fixed point $A$ about which rotation takes place, then

$$\mathbf{V}_{A1} = \mathbf{0} \tag{7.23}$$

Likewise for body 2.

## 7.2 ENERGY LOSS

The importance of examining the kinetic loss $T_L$ during an impact and its bearing on the choice of realistic values of the impact ratios has been established in past chapters. Energy loss should always be calculated for each solution of the system equations. The analytical complexity of the three-dimensional problem does not permit the development of general expressions for $T_L$.

## 7.3 EXAMPLE OF A RIGID PENDULUM

A rigid spherical pendulum whose tip strikes a flat, massive surface is a three-dimensional problem. A velocity constraint on the rigid body must be used to represent the pivot point, so the equations must be revised. Figure 7.3 shows the pendulum and associated coordinates. The $(x, y,$

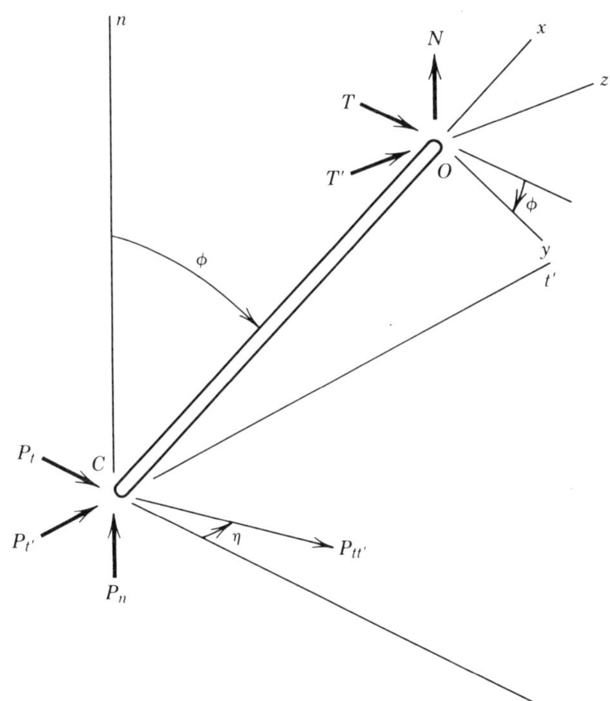


**Figure 7.3.** Rigid spherical pendulum pivoted at $O$; contact point at $C$.

$z$) axes are body axes, where the $x$ axis lies along the pendulum. The $n$ axis is normal to the surface. The general three-dimensional equations presented earlier could be used and constraint equations added; however, the equations are much simpler if rotation about point $O$ is taken into account right from the start. The system's rotational energy about the $x$ axis is neglected and the corresponding angular velocities $\omega_x$ and $\Omega_x$ are assumed to be zero.

The concept of restitution, Eq. 7.6 provides

$$\mathbf{\Omega} \times \mathbf{d} \cdot \mathbf{n} = -e\boldsymbol{\omega} \times \mathbf{d} \cdot \mathbf{n} \tag{7.24}$$

The resultant tangential impulse

$$P_{tt'} = \sqrt{P_t^2 + P_{t'}^2} = \mu P_n \tag{7.25}$$

where $P_t = \mu_t P_n = \mu \cos \eta \, P_n$ and $P_{t'} = \mu_{t'} P_n = \mu \sin \eta \, P_n$. Equating changes in angular momentum to the moment of the impulses about the fixed point $O$ gives

$$I_y(\Omega_y - \omega_y) = d\mu_{t'} P_n \tag{7.26}$$

and

$$I_z(\Omega_z - \omega_z) = d(\sin \phi - \mu_t \cos \phi) P_n \tag{7.27}$$

The system has two degrees-of-freedom corresponding to the rotational velocities about the $y$ and $z$ axis, so there are four unknowns: $\Omega_y$, $\Omega_z$, $P_n$, and $P_{tt'}$. Equation 7.24 simplifies to

$$\Omega_z = -e\omega_z \tag{7.28}$$

Substitution of this into Eq. 7.27 gives

$$P_n = -\frac{(1 + e) I_z \omega_z}{d(\sin \phi - \mu_t \cos \phi)} \tag{7.29}$$

This, in turn, makes Eq. 7.26 become

$$\Omega_y = \omega_y - \mu_{t'} \frac{(1 + e)\omega_z}{(\sin \phi - \mu_t \cos \phi)} \frac{I_z}{I_y} \tag{7.30}$$

Relative motion of the pendulum's tip over the surface (i.e., sliding) can correspond only to $\Omega_y$ and not $\Omega_z$, so the resultant tangential impulse is such that $\eta = \pi/2$. The system's energy loss is

$$T_L = \tfrac{1}{2} I_z \omega_z^2 (1 + e) \left[ (1 - e) + \frac{2\mu}{\sin\phi} \frac{\omega_y}{\omega_z} - \mu^2 \frac{(1 + e) I_z}{\sin^2 \phi \, I_y} \right] \quad (7.31)$$

Setting to zero the derivative of $T_L$ with respect to $\mu$ gives

$$\mu_m = \frac{\sin\phi}{1 + e} \frac{I_y \omega_y}{I_z \omega_z} \quad (7.32)$$

from which for $e = 0$, $\mu_{\max} = I_y \omega_y \sin\phi / I_z \omega_z$. Using Eq. 7.32 in Eq. 7.30 gives

$$\Omega_y = \omega_y (1 - \mu / \mu_m) \quad (7.33)$$

Equations 7.28 and 7.33 provide the solution of the rigid pendulum impact problem. The tangential tip impulse is

$$P_{tt'} = \mu P_n = -\frac{\mu I_y \omega_y}{\mu_m d} \quad (7.34)$$

For the special case of a planar pendulum $\omega_y = 0$ and from Eq. 7.32, $\mu_{\max} = 0$. Here, because the pendulum is rigid and has only one degree of freedom, no tangential impulse can develop.

## 7.4 CLOSURE ON THREE-DIMENSIONAL IMPACT

Although a three-dimensional theory has been presented, the dynamicist must proceed with due caution with regard to realism of solutions. It is not difficult to set up system equations and solve them, but the presence of frictional effects or other mechanisms such as torsional restitution or both that control tangential and rotational impulses can make interpretation of results difficult, if not impossible. It may not be possible to predict if and when certain velocity components end prior to separation. In some cases, it may be wiser to seek a solution by integration of the differential equations of motion.

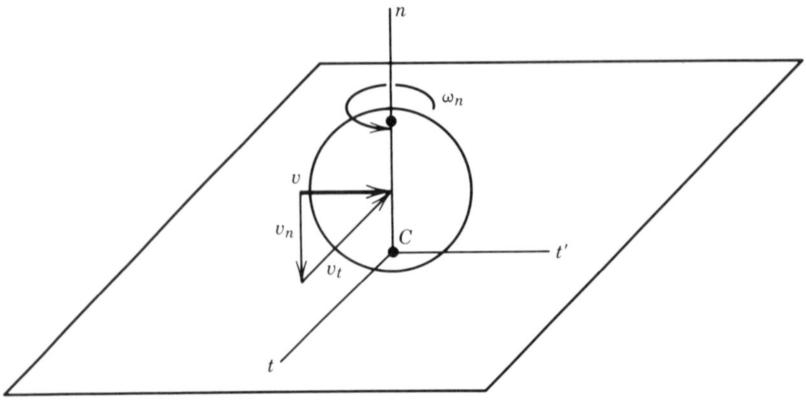


**Figure 7.4.** Spinning sphere.

To clarify this, consider an example of the conceptually simple impact of a spinning sphere against a flat barrier. See Fig. 7.4. To simplify the discussion let $v_{t'} = 0$ and $\omega_t = \omega_{t'} = 0$. A wide range of values for the initial velocities $v_n$, $v_t$, and $\omega_n$ are possible, which for different ranges of sphere and barrier materials imply a wide range of contact durations. For example, a high-velocity glancing impact ($v_t >> v_n$) for hard materials will have a shorter duration of contact than for a largely normal impact ($v_n >> v_t$) of softer materials. Consider that independently $\omega_n$ can be large or small, and it becomes apparent that determination of whether or not this spinning velocity goes to zero prior to separation is not a simple problem. Embed this same decision into a problem with two, nonspherical bodies with nonzero values for all initial velocity components, and the problem may be virtually impossible. Additional research needs to be done on this problem. For one approach see Refs. 3 and 4.

Based on energy considerations, recent research results by W. J. Stronge[5] indicate that for noncentral (eccentric) collisions in the presence of a tangential impulse, with no moment impulse and when relative tangential motion ends or reverses before separation, a definition of the coefficient of restitution different than used in this book is more consistent. His definition is the ratio of the rebound work done by the normal impulse to the corresponding work during approach. Though the differently defined coefficients possess an equivalence, the bounds on the impulse ratio $\mu$ are not independent of bounds on the kinematic coefficient $e$ and so that the expressions for $\mu_c$ such as Eq. 5.46 and those discussed in Section 6.2 should be viewed as estimates in those cases. In light of the assumption for impact theory that angular rotations of rigid bodies during contact are zero, any differences in practice are likely to be small. Exact energy expressions for simultaneous bounds on $e$ and $\mu$ have yet to be established. When a moment impulse exists, simultaneous bounds on $e$, $\mu$ and $e_m$ have never been investigated; this topic requires additional research.

## REFERENCES


1. Rothbart, H. A. (Ed.), *Mechanical Design and Synthesis Handbook*, 2nd ed., McGraw-Hill, New York, 1985.
2. Spotts, M. F., *Design of Machine Elements*, Prentice-Hall, Englewood Cliffs, NJ, 1985.
3. Keller, J. B., "Impact with Friction," *J. Appl. Mech.*, **53,** 1, 1986.
4. Keller, J. B., "Impact with an Impulsive Frictional Moment," Brief Note, *J. Appl. Mech.*, **54,** 1–4, 1987.
5. Stronge, W. J., "Rigid Body Collisions with Friction," *Proc. R. Soc. Lond.*, **A 430,** 169, 1990.

# CHAPTER 8

# PLANAR IMPACT OF LINKAGES AND ARTICULATED RIGID BODIES

To this point, collisions between pairs of point masses and collisions between pairs of rigid bodies or between a rigid body and a barrier have been considered. Many machine components and devices consist of interconnected rigid bodies. Figure 8.1 illustrates a set of general, idealized pin-connected rigid bodies in contact at point $C$ between bodies $L_m$ and $R_i$. Impacts of such chains of rigid bodies are now considered. Certain idealized assumptions are made. The connections between bodies are considered to be ideal. That is, they contain zero clearance and are lossless with no rotational friction and with perfect restitution normal to the pins. It is further assumed that the conditions of impact are felt instantaneously and simultaneously by all rigid bodies involved. If significant joint clearances exist, then the impacts would be sequential and the methods already covered can be applied in a sequence of impacts of adjacent pairs of rigid bodies. In some cases, it may be necessary to perform analyses based on both conditions, simultaneous and sequential impact, in order to determine the effects of the different assumptions. Other approaches can be taken to the problem of impact of multibody systems. As with any other impact problem, the effects of body flexibility cannot always be ignored. The corresponding methods, although tending to be complicated,[1] are often necessary.

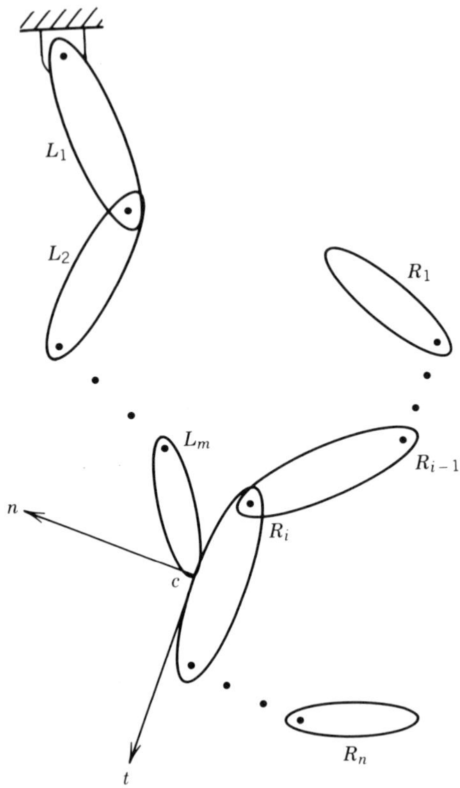


**Figure 8.1.** Impact of body chain $L$ with body chain $R$.

A considerable amount of research has been done on the problem of formulation of the dynamical equations of motion (differential equations of motion) for body chains.[2,3] One of the major objectives of the problem is to obtain the smallest number and simplest set of equations. Direct application of Newton's second law yields a set of equations that explicitly contains the forces of interaction between the rigid bodies. Energy principles such as Lagrange equations yield equations of motion that do not contain these internal forces and moments but can require a significant amount of differentiation. Another part of the problem involves the treatment of constraints. Dynamical constraints can involve prescribed velocities or displacements or both at one or more points of the rigid bodies. In a dynamical context, the equations of motion are second-order differential equations; the constraints usually are first order or possibly algebraic. This provides a mixed set of system dynamical equations. For impact system equations as treated here, the existence of constraints causes little or no difficulty, since the system equations already are algebraic in the velocities; consequently, velocity constraints are automatically compatible. Therefore, the presence of constraints increases the number of equations but does not change their character or form.

Elimination of the internal forces of reaction in the dynamical problem is analogous to elimination of internal impulse variables from the impact system equations. In the earlier chapters, the system equations included the impulses $P_n$ and $P_t$. For planar problem of two masses, the system equations were simple enough so that these impulse variables could be eliminated prior to obtaining the solution equations. With multibody systems, elimination of additional internal impulse velocities can be impractical and Lagrange equations for impact have a distinct advantage. When necessary, internal impulses can be calculated easily after the solution is obtained. Examples of both Newtonian and Lagrangian approaches are presented, however, since the use of computer solutions makes either approach practical.

During an impact, physical disturbances such as deformations and stresses travel through materials as waves at finite speeds. The assumption of simultaneity is not likely to be valid for systems of numerous bodies or highly flexible bodies. The assumption of small changes in position and configuration during the duration of contact may be violated as well. In addition these effects can depend on the configuration of the body chains. Consequently, the number of interconnected bodies to which the following methods can be applied in practice is not likely to be very large.

A body chain can contain loops that consist of a member in contact with more than two others. An impact can occur with one of the loop members or away from a loop. Furthermore, an impact can occur between members of a single body chain (which implies the formation of a loop during the contact duration). These situations are not specifically illustrated here. For these cases, the impact mechanics is no different than what is discussed; the major differences are geometric.

System equations for both the Lagrangian and Newtonian formulations are presented in a rather general form for a collision between two members each from a pair of multilink systems. The Lagrangian formulation involves the use of the principle of virtual work to compute generalized impulses. A reader unfamiliar with these concepts is referred to other texts on dynamics; these methods will not be developed here. An example of the tip of a double pendulum striking a rigid surface will be examined in detail to illustrate that the kinetic energy loss again plays an important role in proper selection of coefficient values. Results are presented of how a body chain impact model can be used with computer simulation models of robotic manipulators. Finally, an application to collisions involving automobile–trailer combinations or tractor semitrailers is covered.

## 8.1 SYSTEM EQUATIONS FOR IMPACT OF TWO MULTIBODY SYSTEMS

It is well known that Lagrange's equations can be used to furnish the differential equations of motion of complex mechanical systems.[2-5] For collisions, however, an integrated form of Lagrange's equations is needed. Goldsmith[6] shows that these can be written as

$$\Delta\left(\frac{\partial T}{\partial \dot{q}_i}\right) = P_i, \qquad i = 1, 2, \ldots, N \tag{8.1}$$

where $\dot{q}_i$ is the velocity associated with the generalized coordinate $q_i$, $T$ is the kinetic energy of the system, $P_i$ is the generalized impulse associated with the coordinate $q_i$, $N$ is the number of independent coordinates (degrees of freedom), and $\Delta$ indicates a change over the time interval $\tau_2 - \tau_1$. Lagrange's equations in their classical form for a dynamical system contain the potential energy of the system, whereas Eq. 8.1 does not. This is due to the fundamental assumption of rigid body impact theory that all position coordinates and, consequently, the potential energy remain unchanged during the contact duration. To apply Eq. 8.1 to a multibody system such as shown in Fig. 8.1 it is necessary to

1. establish generalized coordinates of bodies $L_1, L_2, \ldots, L_m$, say $N_L$ of them;
2. determine the kinetic energy expression for bodies $L_1, L_2, \ldots, L_m$ in terms of the coordinates and their velocities;
3. determine the generalized impulses $P_i$ due to $P_n$, $P_t$, and $M$ at point $C$ associated with each coordinate $q_i$ (using virtual work);

4. use all of these expressions to carry out the operations indicated by Eq. 8.1;
5. repeat this process for bodies $R_1, R_2, \ldots, R_n$: $N_R$ coordinates.

This will produce $N_L$ and $N_R$ algebraic equations. There are $N_L + N_R$ unknown final velocities. The impulse variables $P_n$, $P_t$, and $M$ add three more unknowns for a total of $N_L + N_R + 3$. However, three additional equations follow from the definitions of the coefficients $e$, $e_m$, and $\mu$. The resulting $N_L + N_R + 3$ equations are the system equations. Note that they do not contain the internal impulse components at each of the joints between the members of system $L$ and those of system $R$.

### 8.1.1 Application to a Two-Link Chain and Free Mass

Figure 8.2 shows a simple chain composed of two long slender members of mass $m_1$ and $m_2$ with $m_1$ pinned at point $O$. The tip of $m_2$ impacts a free mass $m_3$. The mass centers of $m_1$ and $m_2$ are taken midway between their ends but are otherwise arbitrarily shaped. Equation 8.1 will be used to develop the system equations for this system. The total kinetic energy

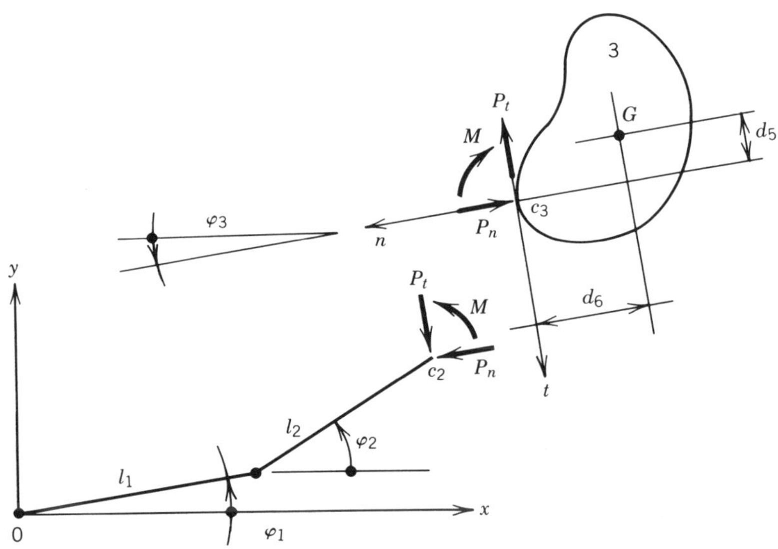


**Figure 8.2.** Collision of an unrestrained rigid body with a two-link chain.

for the two members can be written as

$$T_{12} = \tfrac{1}{2}(m_1 l_1^2/4 + m_2 l_1^2 + I_1)\dot{\phi}_1^2 + \tfrac{1}{2}(m_2 l_2^2/4 + I_2)\dot{\phi}_2^2$$
$$+ \tfrac{1}{2}m_2 l_1 l_2 \cos(\phi_2 - \phi_1)\dot{\phi}_1\dot{\phi}_2 \tag{8.2}$$

For mass $m_3$, the kinetic energy is

$$T_3 = \tfrac{1}{2}m_3(\dot{n}_G^2 + \dot{t}_G^2) + \tfrac{1}{2}I_3\dot{\phi}_3^2 \tag{8.3}$$

The generalized coordinates are $q_1 = \phi_1$, $q_2$, $= \phi_2$, $q_3 = n_G$, $q_4 = t_G$, and $q_5 = \phi_3$, where $\phi_1$, $\phi_2$, and $\phi_3$ are the angular coordinates of masses $m_1$, $m_2$, and $m_3$, respectively, with respect to the fixed $x$-$y$ coordinate system; $n_G$ and $t_G$ are mass center coordinates of mass 3. Dots indicate derivatives with respect to time $\tau$. The problem has five unknown final velocity components, $\Omega_1$, $\Omega_2$, $V_{3n}$, $V_{3t}$, and $\Omega_3$, and three unknown impulse components, $P_n$, $P_t$, and $M$. Notation is such that $\Omega_1 = \dot{\phi}_1(\tau_2)$, $V_{3n} = \dot{n}_G(\tau_2)$, and so forth. As is customary, initial velocities are in lower case. Five equations are obtained from Eq. 8.1 by letting $q_i$ successively be $\phi_1$, $\phi_2$, $n_G$, $t_G$, and $\phi_3$. These are, respectively,

$$(m_1 l_1^2/4 + m_2 l_1^2 + I_1)(\Omega_1 - \omega_1) + \tfrac{1}{2}m_2 l_1 l_2 \cos(\phi_2 - \phi_1)(\Omega_2 - \omega_2)$$
$$= l_1 \sin(\phi_1 - \phi_3)P_n - l_1 \cos(\phi_1 - \phi_3)P_t \tag{8.4}$$

$$\tfrac{1}{2}m_2 l_1 l_2 \cos(\phi_2 - \phi_1)(\Omega_1 - \omega_1) + (m_2 l_2^2/4 + I_2)(\Omega_2 - \omega_2)$$
$$= l_2 \sin(\phi_2 - \phi_3)P_n - l_2 \cos(\phi_2 - \phi_3)P_t + M \tag{8.5}$$

$$m_3(V_{3n} - v_{3n}) = -P_n \tag{8.6}$$

$$m_3(V_{3t} - v_{3t}) = -P_t \tag{8.7}$$

$$I_3(\Omega_3 - \omega_3) = -M + d_5 P_n - d_6 P_t \tag{8.8}$$

where $\Delta\dot{\phi}_i = \Omega_i - \omega_i$, $\Delta\dot{n}_G = V_{3n} - v_{3n}$, and $\Delta\dot{t}_G = V_{3t} - v_{3t}$. The generalized momenta (right-hand sides of Eq. 8.10) were found using the method of virtual work.[7,8]. The three remaining system equations are found from the definition of the coefficients $e$, $\mu$, and $e_m$. These are

$$\begin{aligned}
&[l_1 \sin(\phi_1 - \phi_3)\,\Omega_1 + l_2 \sin(\phi_2 - \phi_3)\,\Omega_2] - (V_{3n} - d_5\,\Omega_3) \\
&\quad = -e\left\{[l_1 \sin(\phi_1 - \phi_3)\omega_1 + l_2 \sin(\phi_2 - \phi_3)\omega_2] \right. \\
&\quad\quad \left. - (v_{3n} - d_5\,\omega_3)\right\}
\end{aligned} \tag{8.9}$$

$$P_t = \mu P_n \tag{8.10}$$

and

$$e_m M = -(1 + e_m)\bar{I}(\Omega_3 - \Omega_2) \tag{8.11}$$

where $\bar{I} = I_2 I_3/(I_2 + I_3)$.

### ■ *Example 8.1*

PROBLEM: A two-link manipulator has masses $m_1 = 2m$ and $m_2 = m$, kg, with inertias $I_1 = m/3$ and $I_2 = m/9$, kg-m$^2$. While moving with velocities $\omega_1 = 1$ and $\omega_2 = 3$ rad/s, it grasps rigidly a stationary load of mass $m_3 = m/10$ kg and inertia $I_3 = m/30$ kg-m$^2$. The quantities $\phi_1 = 0$, $\phi_2 = 0$, $\phi_3 = 0$, $l_1 = 1$, $l_2 = 0.5$ m, $d_5 = 0$ and $d_6 = 0.1$ m.

Determine the pertinent velocities of the manipulator, the contact point, and the mass center of the load; determine the impulses $P_n$, $P_t$, and $M$.

*Solution:* Since the load is grasped rigidly, $e = 0$, $\mu = \mu_0$, and $e_m = 0$. For the configuration with $\phi_1 = \phi_2 = \phi_3 = 0$, Eq. 8.9 shows that $V_{3n} = 0$. Equation 8.6 then shows that $P_n = 0$. For this reason, Eq. 8.10 must be replaced by an equation that constrains the tangential velocities of the contact point in bodies 2 and 3 to be the same. Equations 8.4, 8.5, 8.7, 8.8, 8.10 (replaced), and 8.11 become, respectively,

$$22m\Omega_1 + 3m\Omega_2 + 12P_t = 31m \tag{a}$$

$$36m\Omega_1 + 25m\Omega_2 + 72P_t - 144M = 111m \tag{b}$$

$$mV_{3t} + 10P_t = 0 \tag{c}$$

$$m\Omega_3 + 3P_t + 30M = 0 \tag{d}$$

$$V_{3t} + \Omega_1 + 0.5\Omega_2 + 0.1\Omega_3 = 0 \tag{e}$$

$$\Omega_2 - \Omega_3 = 0 \tag{f}$$

These equations can be solved using the method of substitution to yield the following results:

| Initial | Final | Impulses |
|---|---|---|
| $\omega_1 = 1$ rad/s | $\Omega_1 = 0.9176$ rad/s | $P_n = 0$ |
| $\omega_2 = 3$ rad/s | $\Omega_2 = -0.0134$ rad/s | $P_t = 0.0926m$ N-s |
| $\omega_3 = 0$ rad/s | $\Omega_3 = -0.0134$ rad/s | $M = -0.0088m$ N-s-m |
| $v_{3n} = 0$ m/s | $v_{3n} = 0$ | |
| $v_{3t} = 0$ m/s | $v_{3t} = -0.9256$ m/s | |
| $v_{cn} = 0$ | $v_{cn} = 0$ | |
| $v_{ct} = 0$ | $V_{ct} = -0.9109$ m/s | ■ |

### 8.1.2 Application: Impact of a Two-Link Body Chain (Double Pendulum) with a Rigid Surface

A solution of the double-pendulum problem is presented here for a range of parameter values in order to investigate the concept of the critical impulse ratio. Imagine that mass $m_3$ of Fig. 8.2 forms a massive, rigid flat stationary surface (barrier) making an angle $\phi_3 = \alpha$ to the $y$ axis. A point impact with no moment is assumed, so that $e_m = -1$. Equations 8.4, 8.5, 8.9, and 8.10 are the system equations. The unknowns are $\Omega_1$, $\Omega_2$, $P_n$, and $P_t$. All of the velocity components of body 3 are zero in Eq. 8.9 for a stationary surface. The physical process tangent to the surface will be taken to be that of Coulomb friction with a coefficient $f$. This problem is adapted from one presented in Ref. 9. In that problem as here, $\phi_1 = 20°$, $\phi_2 = 30°$, $\dot{\phi}_1 = \omega_1 = -0.1$ rad/s, and $\dot{\phi}_2 = \omega_2 = -0.2$ rad/s. The rods are uniform with $m_1 = m_2 = 3$ kg and $l_1 = l_2 = 2$ m. The coefficient of restitution will be varied between $0 \leq e \leq 1$ and the friction coefficient* between $0 \leq f \leq 1$. The final values of the rods' angular velocities will be found for combinations of these parameters.

*In Ref. 9, static and dynamic coefficients of friction are supplied. The static coefficient generally does not apply to impact; see Chapter 3.

The solution equations are attainable in a closed form for this problem. Equations 8.4 and 8.5 can be used to solve simultaneously for $P_n$ and $P_t$. Substitution of these into Eq. 8.10 leaves Eqs. 8.9 and 8.10 as two equations solvable for $\Omega_1$ and $\Omega_2$. Ultimately, the following is obtained:

$$\Omega_1 = \omega_1 - (1 + e)\frac{l_1 s_1 \omega_1 + l_2 s_2 \omega_2}{(l_1 s_1 + l_2 s_2)A} \tag{8.12}$$

and

$$\Omega_2 = \omega_2 + (\Omega_1 - \omega_1)A \tag{8.13}$$

where

$$A = \frac{(I_{11}/l_1)(s_2 - \mu c_2) - (I_{21}/l_2)(s_1 - \mu c_1)}{(I_{22}/l_2)\,(s_1 - \mu c_1) - (I_{12}/l_1)(s_2 - \mu c_2)} \tag{8.14}$$

$$I_{11} = I_1 + \tfrac{1}{4} m_1 l_1^2 + m_2 l_1^2 \tag{8.15}$$

$$I_{12} = I_{21} = \tfrac{1}{2} m_2 l_1 l_2 \cos(\phi_2 - \phi_1) \tag{8.16}$$

$$I_{22} = I_2 + \tfrac{1}{4} m l_2^2 \tag{8.17}$$

and

$$\begin{aligned} s_1 &= \sin(\phi_1 - \alpha), \qquad & s_2 &= \sin(\phi_2 - \alpha) \\ c_1 &= \cos(\phi_1 - \alpha), \qquad & c_2 &= \cos(\phi_2 - \alpha) \end{aligned} \tag{8.18}$$

A singularity occurs in the solution for $P_n$ and $P_t$ for $\phi_1 = \phi_2$. For this condition the impulse at the tip of the rod can only be perpendicular to the rod since the change in momentum along the rod must be zero. A singularity also occurs when the denominator of $A$ of Eq. 8.14 is zero. In both cases solutions can be found by appropriate manipulation of equations. Singularities do not appear for the numerical parameters in this example.

As seen throughout this book, the system equations are solved by treating the impulse ratio $\mu$ as a known constant, yet the actual physical parameter at the surface is the friction coefficient $f$. Examination of the system's kinetic energy loss is used to uncover the relationship between $\mu$ and $f$. The procedure can be summarized as follows:

1. Calculate the final tangential velocity at the contact point for the pertinent value of the coefficient of restitution and for a frictionless impact, $f = \mu = 0$.
2. Incrementally increase the magnitude of $\mu$ from 0 to a value of $\pm f$ (the appropriate sign is such that the tangential impulse opposes the final tangential velocity to be zero ($\mu_0$) and the kinetic energy loss to become negative ($\mu_T$)).
3. The smallest in magnitude of the three quantities $f$, $\mu_0$, and $\mu_T$ is used for $\mu$ in the solution equations to obtain the appropriate solution. The smaller (in magnitude) of $\mu_0$ and $\mu_T$ is the critical value, $\mu_c$.

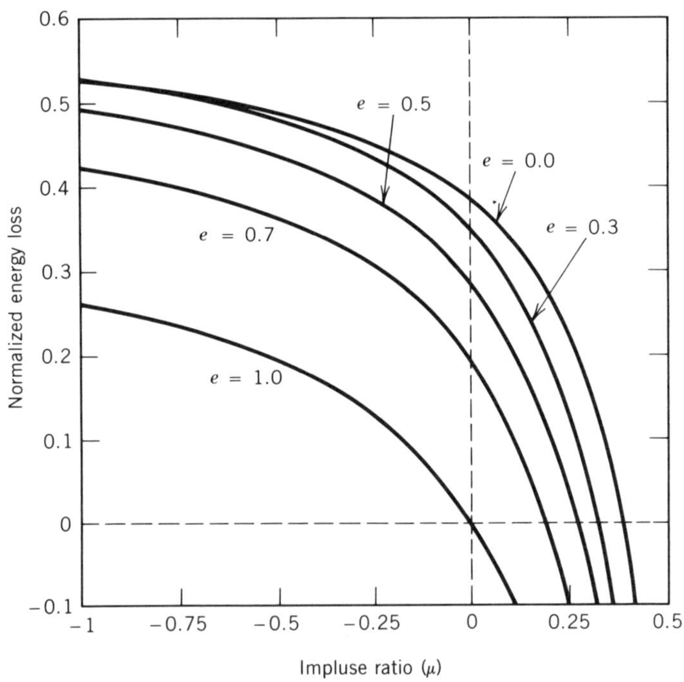


**Figure 8.3.** Energy loss, tip impact of a double pendulum with a rigid surface.

This procedure is now followed for the impact of the tip of the double link. Figure 8.3 shows the normalized energy loss for several values of $e$. Table 8.1 shows the solutions for $e = 1$ and $-1 \leq \mu = 1$. The final tangential tip velocity, $V_{ct}$, for $f = \mu = 0$ is $-0.275$ m/s. Addition of a positive tangential impulse $P_t$ with a value of $\mu > 0$ causes the final tangential tip velocity at point $C$ to decrease, but this can be done only with an increase in kinetic energy of the system (negative energy loss). That is, $\mu_T = 0$ here. Using a negative value of $\mu$ causes the final tip velocity to be greater (in magnitude) than the frictionless case. Consequently, a value of $\mu = 0$ provides the final velocities for an elastic impact of this double pendulum for $e = 1$, regardless of the actual value of the friction coefficient. Table 8.2 shows the solutions for $e = 0.5$. For $\mu = 0$, the final tangential tip velocity is $-0.072$ with an energy loss of 28.9%. Since the final tangential velocity is negative, a positive

**TABLE 8.1 Double Pendulum Problem; Solution of System Equations, $e = 1.0$**

| $\mu$ | $\Omega_1$ | $\Omega_2$ | $V_{cn}$ | $V_{ct}$ | $P_n$ | $P_t$ | $T_L$ |
|---|---|---|---|---|---|---|---|
| −1.000 | −0.229 | 0.425 | 0.268 | −0.306 | 0.636 | −0.636 | 0.261 |
| −0.900 | −0.231 | 0.426 | 0.268 | −0.305 | 0.677 | −0.609 | 0.252 |
| −0.800 | −0.233 | 0.428 | 0.268 | −0.303 | 0.723 | −0.579 | 0.241 |
| −0.700 | −0.235 | 0.429 | 0.268 | −0.301 | 0.776 | −0.543 | 0.228 |
| −0.600 | −0.238 | 0.431 | 0.268 | −0.299 | 0.838 | −0.503 | 0.212 |
| −0.500 | −0.242 | 0.434 | 0.268 | −0.297 | 0.910 | −0.455 | 0.194 |
| −0.400 | −0.246 | 0.437 | 0.268 | −0.294 | 0.995 | −0.398 | 0.172 |
| −0.300 | −0.251 | 0.440 | 0.268 | −0.291 | 1.099 | −0.330 | 0.144 |
| −0.200 | −0.257 | 0.444 | 0.268 | −0.287 | 1.226 | −0.245 | 0.109 |
| −0.100 | −0.264 | 0.449 | 0.268 | −0.281 | 1.386 | −0.139 | 0.063 |
| 0.000 | −0.274 | 0.456 | 0.268 | −0.275 | 1.595 | 0.000 | 0.000 |
| 0.100 | −0.287 | 0.465 | 0.268 | −0.265 | 1.878 | 0.188 | −0.091 |
| 0.200 | −0.306 | 0.478 | 0.268 | −0.252 | 2.283 | 0.457 | −0.232 |
| 0.300 | −0.336 | 0.498 | 0.268 | −0.232 | 2.911 | 0.873 | −0.475 |
| 0.400 | −0.388 | 0.534 | 0.268 | −0.195 | 4.016 | 1.606 | −0.979 |
| 0.500 | −0.504 | 0.613 | 0.268 | −0.115 | 6.470 | 3.235 | −2.438 |
| 0.600 | −0.983 | 0.941 | 0.268 | 0.218 | 16.614 | 9.985 | −13.494 |
| 0.700 | 1.171 | −0.533 | 0.268 | −1.278 | −29.092 | −20.365 | −27.234 |
| 0.800 | 0.167 | 0.154 | 0.268 | −0.581 | −7.762 | −6.209 | −0.517 |
| 0.900 | 0.012 | 0.260 | 0.268 | −0.473 | −4.478 | −4.030 | 0.443 |
| 1.000 | −0.051 | 0.303 | 0.268 | −0.430 | −3.147 | −3.147 | 0.592 |

**TABLE 8.2 Double Pendulum Problem; Solution of System Equations, $e = 0.5$**

| $\mu$ | $\Omega_1$ | $\Omega_2$ | $V_{cn}$ | $V_{ct}$ | $P_n$ | $P_t$ | $T_L$ |
|---|---|---|---|---|---|---|---|
| −1.000 | −0.197 | 0.269 | 0.134 | −0.096 | 0.477 | −0.477 | 0.491 |
| −0.900 | −0.198 | 0.270 | 0.134 | −0.095 | 0.508 | −0.457 | 0.483 |
| −0.800 | −0.200 | 0.271 | 0.134 | −0.094 | 0.542 | −0.434 | 0.475 |
| −0.700 | −0.202 | 0.272 | 0.134 | −0.092 | 0.582 | −0.408 | 0.464 |
| −0.600 | −0.204 | 0.274 | 0.134 | −0.091 | 0.628 | −0.377 | 0.452 |
| −0.500 | −0.206 | 0.275 | 0.134 | 0.089 | 0.682 | −0.341 | 0.438 |
| −0.400 | −0.209 | 0.277 | 0.134 | −0.087 | 0.746 | −0.299 | 0.420 |
| −0.300 | −0.213 | 0.280 | 0.134 | −0.085 | 0.824 | −0.247 | 0.399 |
| −0.200 | −0.217 | 0.283 | 0.134 | −0.081 | 0.919 | −0.184 | 0.371 |
| −0.100 | −0.223 | 0.287 | 0.134 | −0.077 | 1.040 | −0.104 | 0.336 |
| 0.000 | −0.231 | 0.292 | 0.134 | −0.072 | 1.196 | 0.000 | 0.289 |
| 0.100 | −0.240 | 0.299 | 0.134 | −0.065 | 1.409 | 0.141 | 0.221 |
| 0.200 | −0.255 | 0.309 | 0.134 | −0.055 | 1.712 | 0.342 | 0.118 |
| 0.300 | −0.277 | 0.324 | 0.134 | −0.040 | 2.183 | 0.655 | −0.055 |
| 0.400 | −0.316 | 0.350 | 0.134 | −0.013 | 3.012 | 1.205 | −0.403 |
| 0.500 | −0.403 | 0.410 | 0.134 | 0.047 | 4.853 | 2.426 | −1.366 |
| 0.600 | −0.762 | 0.655 | 0.134 | 0.297 | 12.481 | 7.488 | −8.176 |
| 0.700 | 0.853 | −0.450 | 0.134 | −0.825 | −21.819 | −15.273 | −13.248 |
| 0.800 | 0.100 | 0.066 | 0.134 | −0.302 | −5.821 | −4.657 | 0.542 |
| 0.900 | −0.016 | 0.145 | 0.134 | −0.221 | −3.359 | −3.023 | 0.890 |
| 1.000 | −0.063 | 0.177 | 0.134 | −0.189 | −2.360 | −2.360 | 0.897 |

tangential impulse decreases $V_{ct}$. The limiting values of the impulse ratio are $\mu_0 = 0.408$ and $\mu_T = 0.268$. If the friction coefficient between the rod's tip and the surface is $0 \le f \le 0.268$, then $\mu = f$; if it is larger, then the appropriate solution is with $\mu = \mu_c = \mu_T = 0.268$. A negative value of $\mu$ is unrealistic since it causes $V_{ct}$ to be greater in magnitude than the frictionless case.

At this point it is convenient to view the condition $e = 0$ as a kinematic constraint on the final velocities such that the tip cannot leave the surface and must travel only parallel to it. Finally consider the solutions in Table 8.3 for $e = 0$, a perfectly inelastic impact. For $\mu = 0$, $V_{ct} = 0.130$ m/s. Negative values of $\mu$ cause $V_{ct}$ to decrease; positive tangential impulses are not appropriate under these conditions. No negative values of $\mu$ cause $V_{ct}$ to be zero or the energy loss to be negative. Con-

**TABLE 8.3 Double Pendulum Problem; Solution of System Equations, $e$ = 0.0**

| $\mu$ | $\Omega_1$ | $\Omega_2$ | $V_{cn}$ | $V_{ct}$ | $P_n$ | $P_t$ | $T_L$ |
|---|---|---|---|---|---|---|---|
| −1.000 | −0.164 | 0.112 | 0.000 | 0.114 | 0.318 | −0.318 | 0.524 |
| −0.900 | −0.165 | 0.113 | 0.000 | 0.115 | 0.339 | −0.305 | 0.519 |
| −0.800 | −0.166 | 0.114 | 0.000 | 0.116 | 0.362 | −0.289 | 0.512 |
| −0.700 | −0.168 | 0.115 | 0.000 | 0.116 | 0.388 | −0.272 | 0.505 |
| −0.600 | −0.169 | 0.116 | 0.000 | 0.118 | 0.419 | −0.251 | 0.497 |
| −0.500 | −0.171 | 0.117 | 0.000 | 0.119 | 0.455 | −0.227 | 0.486 |
| −0.400 | −0.173 | 0.118 | 0.000 | 0.120 | 0.498 | −0.199 | 0.474 |
| −0.300 | −0.175 | 0.120 | 0.000 | 0.122 | 0.549 | −0.165 | 0.459 |
| −0.200 | −0.178 | 0.122 | 0.000 | 0.124 | 0.613 | −0.123 | 0.441 |
| −0.100 | −0.182 | 0.125 | 0.000 | 0.126 | 0.693 | −0.069 | 0.417 |
| 0.000 | −0.187 | 0.128 | 0.000 | 0.130 | 0.798 | 0.000 | 0.385 |
| 0.100 | −0.194 | 0.132 | 0.000 | 0.135 | 0.939 | 0.094 | 0.340 |
| 0.200 | −0.203 | 0.139 | 0.000 | 0.141 | 1.142 | 0.228 | 0.274 |
| 0.300 | −0.218 | 0.149 | 0.000 | 0.151 | 1.456 | 0.437 | 0.164 |
| 0.400 | −0.244 | 0.167 | 0.000 | 0.169 | 2.008 | 0.803 | −0.047 |
| 0.500 | −0.302 | 0.206 | 0.000 | 0.210 | 3.235 | 1.618 | −0.602 |
| 0.600 | −0.541 | 0.370 | 0.000 | 0.376 | 8.320 | 4.992 | −4.154 |
| 0.700 | 0.536 | −0.366 | 0.000 | −0.372 | −14.546 | −10.182 | −4.046 |
| 0.800 | 0.033 | −0.023 | 0.000 | −0.023 | −3.881 | −3.105 | 0.980 |
| 0.900 | −0.044 | 0.030 | 0.000 | 0.031 | −2.239 | −2.015 | 0.966 |
| 1.000 | −0.075 | 0.052 | 0.000 | 0.052 | −1.573 | −1.573 | 0.900 |

sequently, the proper solution of the impact problem here is for $\mu = -f$, $f \geq 0$.

It is possible to determine (by varying $e$ and generating numerous solutions) that for $\mu = 0$, $V_{ct} = 0$ when the coefficient of restitution $e = 0.322$. For this value of $e$, any value of $\mu \neq 0$ causes the final tangential tip velocity to increase in magnitude. Consequently, for this value of $e$, $\mu_c = 0$ for all values of the surface friction coefficient.

## 8.2 IMPACT OF MANIPULATORS

A method is presented in this section through which the effects of an impact on the motion of manipulators can be computed. It is intended to be used in conjunction with simulation techniques in the design phase of robotic systems. Manipulators or robot arms are computer-controlled machines with articulated components and are programmable to carry out a variety of tasks. The range of power, size, versatility, and configuration is quite broad. Despite their complexity in design and manufacture, as any physical device, their operation must follow physical laws. Various branches of engineering that interrelate diverse subjects such as kinematics, feedback control, hydraulics, digital electronics, and so forth, must be considered. A very narrow view of manipulators is taken here in order to illustrate a technique for studying the effects of impact on manipulator motion. Only planar motion and only revolute joints are considered. Consistent with the remainder of the book, a rigid body approach is taken.

The technique uses the system equations for rigid body chains to represent a manipulator. The system equations are used in conjunction with an independent dynamical simulation of a given manipulator. When an impact occurs, the dynamical simulation is interrupted to furnish a set of initial velocity components for the system equations. The system equations are then solved for a new set of velocities; these are then used as a set of initial conditions for continuation of the simulation. The simulation then continues to determine the path of the manipulator as affected by the velocity change (impact). For multiple impacts, each is treated individually.

Impacts to manipulators can occur under different circumstances. One of the most obvious is an unintended collision with an object or barrier. A great deal of planning is done[10] to prevent such events; yet a good design includes provisions for unintended but foreseeable incidents. Another way in which impacts can occur is when the manipulator while in motion picks up or releases an object. In the classical sense of momentum change, this is an impact. As manipulators are designed to operate at higher speeds, it becomes inefficient and may be unnecessary to bring the end effector to a complete stop when an object is grasped. This type of impact requires simulation in the design stage. Another type of impact is when the manipulator is intended to be an impacter. Although not too common yet, the possibility exists that robots may carry out such tasks as driving nails, compacting powders, breaking objects, and so on.

Robot control is typically categorized into two types, motion control and force control.[11-13] When under motion control, the manipulator is free from contact with other objects. When under force control, the end effector or gripper exerts a force or a moment or both on an object, which may be either in its grasp or which may be fixed (immovable). Motion and force control sometimes occur simultaneously. Not only can the control characteristics change drastically from one mode to the other, but when under force control, the control strategy or scheme is usually dependent on the mechanical properties (elasticity, inertia, and so on) of the end effector and the grasped object. This creates a difficult problem for the control system designer in simulation studies. When a manipulator comes into contact with an object, its simulation not only must switch from motion control to force control, but it may need to introduce or modify drastically its structural flexibility modeling capability. In other words, when moving freely through space the inertia and structural properties of a manipulator may be unimportant to a simulation, but when it comes into contact with another object, this may no longer be true. These concepts may apply whether the contact is developed slowly (no impact) or rapidly.

When a manipulator is relatively rigid, the procedure for modeling impacts introduced here requires little or no changes to a dynamical simulation model. When an impact occurs, it is necessary to interrupt the simulation and transfer existing velocities to the impact model. The impact equations calculate a new set of velocity components. These are passed back to the simulation as a set of initial conditions and it continues. In some situations, this may require some restarting of numerical

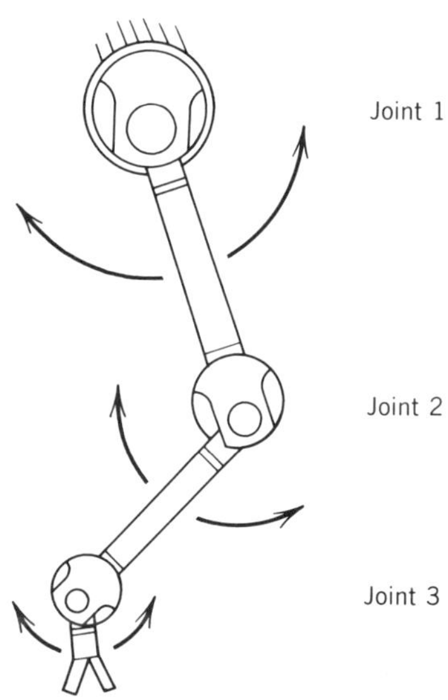


**Figure 8.4.** Three-revolute-jointed manipulator (from Ref. 14, with permission of MIT Press, copyright 1987).

integration schemes, which should prove tractable. All of the physical properties of the contact phenomenon are embedded in the coefficients of restitution and the impulse ratio. These may be unknown and may have to be estimated analytically or determined experimentally. In some cases this technique may not be precise enough to gauge the effects of impact such as in the presence of high flexibility. In other cases it may be sufficient, such as determining the effects of impact on a control strategy. The procedure is described briefly in the next section. The presentation is descriptive and follows from an application to an existing simulation.

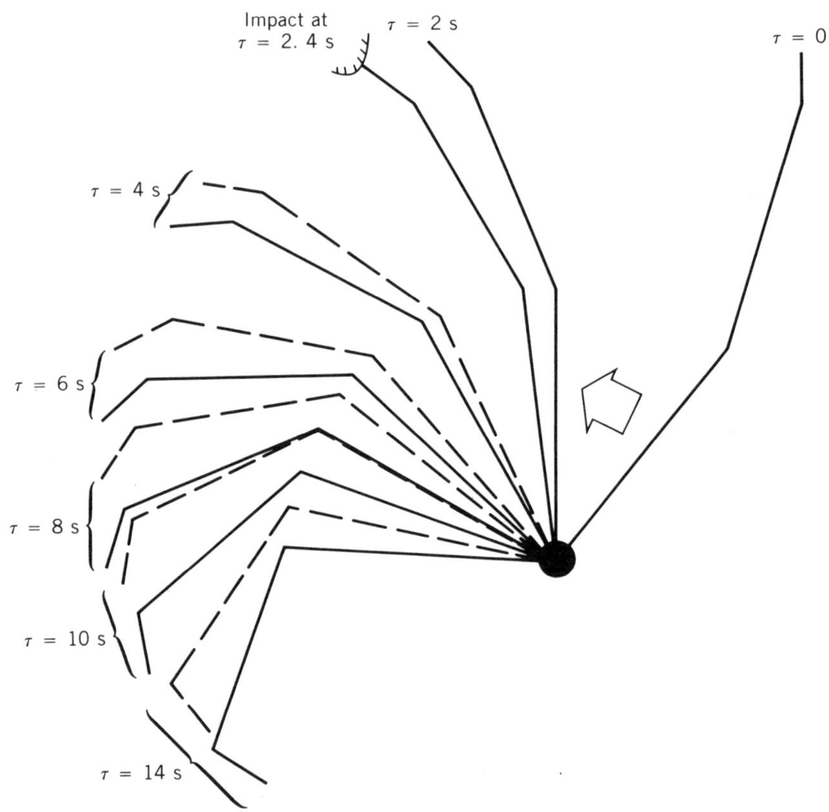


**Figure 8.5.** Motion sequence of the three-jointed manipulator impact, with open-loop control: solid lines are the desired trajectory, while dashed lines are the trajectory with impact.

### 8.2.1 Simulation of Manipulator Impacts

Researchers at Purdue University[14,15] designed and built the three-segment planar manipulator sketched in Fig. 8.4. They also developed a complete set of equations of motion of the system for either open-loop control or closed-loop control. This set of equations was programmed for solution on a personal computer to provide a working simulation, and an impact model was developed and added.[16] Impacts to the end effector were simulated for a wide range of parameter values; a summary is presented in the following paragraphs.

The procedure of introducing an impact into a dynamical simulation worked well. For open-loop control, the impact introduced an eventual

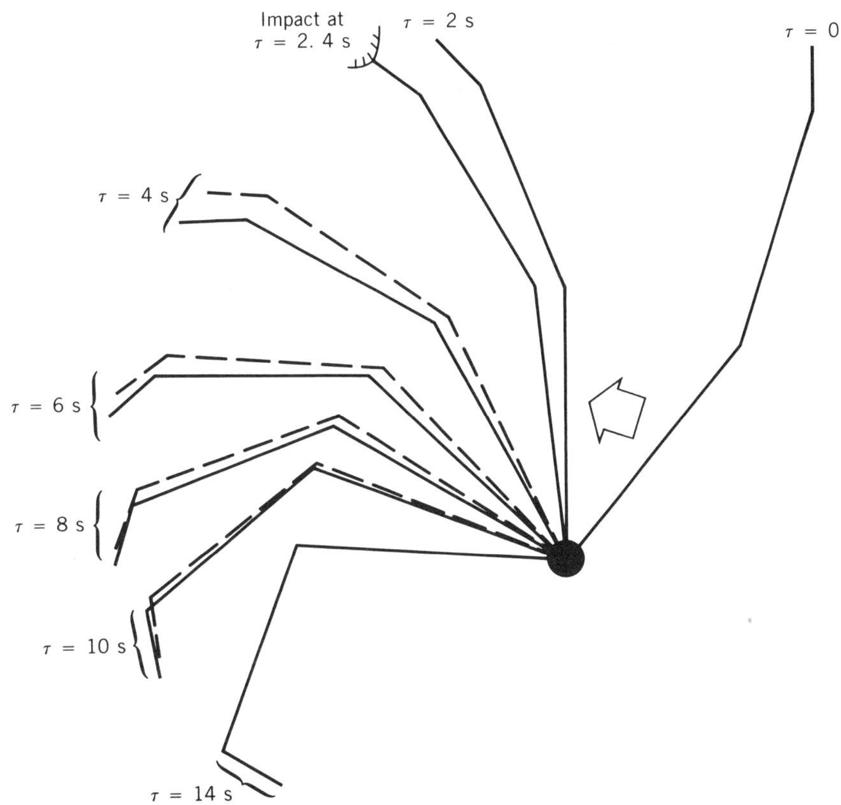


**Figure 8.6.** Motion sequence of the three-jointed manipulator impact, with closed-loop control: solid lines are the desired trajectory, while dashed lines are the trajectory with impact.

displacement and velocity error relative to a specified path. The velocity error disappeared typically within 0.5 s (which depends on the control system gains and system properties), but the displacement errors never disappeared. Figure 8.5 is an example plot of arm positions for a simulation of 14 s of motion with an impact occurring at time $\tau = 2.4$ s. (The displacement deviations in this figure are exaggerated for illustrative purposes.) A similar simulation under closed-loop control (similarly exaggerated) is shown in Fig. 8.6. Here the feedback causes both of the velocity and displacement errors to diminish.

The procedure of interrupting a dynamical system simulation to model an impact of a dynamic system is illustrated here for manipulators. It performs well and is easily adapted to other joint conditions and to three-dimensional motion.

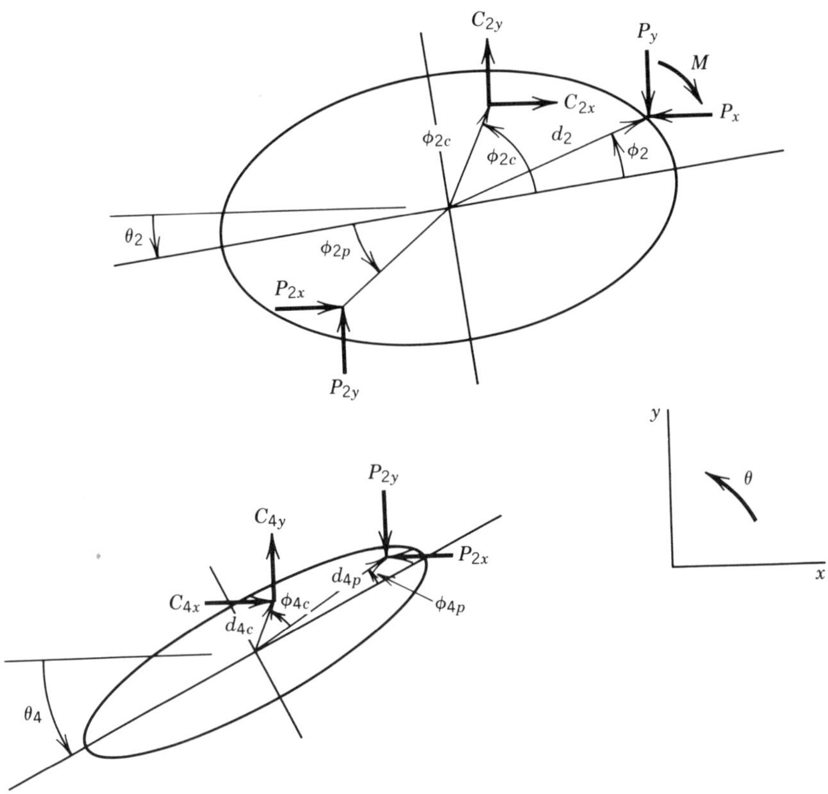


**Figure 8.7.** Free body diagrams of two-member body chains.

## 8.3 IMPACT OF A PAIR OF TWO-MEMBER BODY CHAINS: IMPACT OF ARTICULATED VEHICLES

Earlier in this chapter, the system equations were presented in a form obtainable through the use of Lagrange's equations. Although direct application of Newton's second law in momentum form does not yield as compact a set of equations as Lagrange's, there are cases where they may be desirable. The system equations for a collision of two of each of a pair of pinned bodies is presented here; Fig. 8.7 shows free body diagrams. One pair consists of bodies 1 and 3, the other has bodies 2

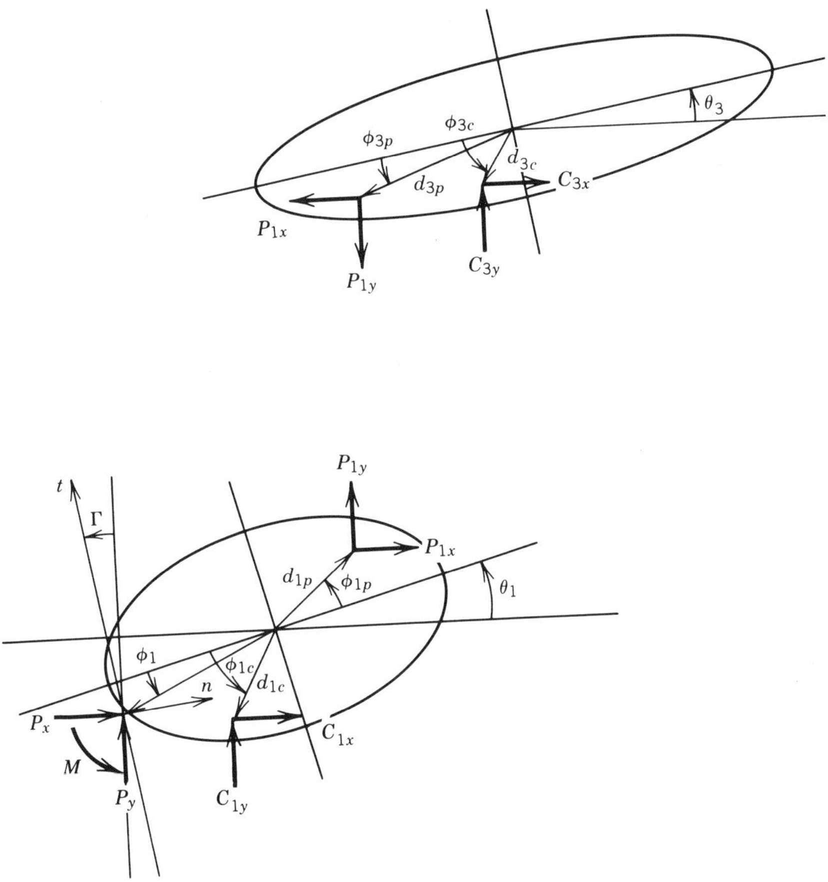


**Figure 8.7.** (*Continued*)

and 4. Each pair is pin-connected and has common impulse components $P_{1x}$, $P_{1y}$ and $P_{2x}$, $P_{2y}$, respectively, at the pins. At the pinned connection in each body pair, the final velocity components must be equal. This must be explicitly enforced and the corresponding equation forms one of the system equations. If a pinned connection does not exist prior to the impact and is established during impact, the initial velocity components are arbitrary. If the bodies are pinned prior to the impact, the initial velocity components must also satisfy the equal velocity constraint at each pinned connection.

Just as the common velocity condition at each pinned connection is associated with an impulse, any other constraint must also. Each body in Fig. 8.7 includes impulse components $C_{ix}$, $C_{iy}$ (and corresponding coordinates $d_{ic}$, $\phi_{ic}$) to represent an arbitrary constraint. For example, if body 3 is pinned to a fixed point, then the impulse components $C_{3x}$ and $C_{3y}$ form part of the system of unknowns and equations expressing zero velocity components at the pinned joint must be introduced as part of the system equations. Of course, the constraints are not limited to pinned connections. A point of a body may be required to have only a $y$ velocity component (because of continual contact with a frictionless rigid barrier) for example. In this case $V_{3Cx} = 0$ and $C_{3x}$ is an unknown.

The number of unknowns depends on the number of constraints. With no external constraints, the unknowns include 12 final velocity components and seven impulse components. The system equations are comprised of 12 equations relating linear and angular impulse (Newton's second law), three equations from the impact coefficients, and four equations of internal constraints from the pinned connections.

For each body, linear impulse and momentum are related by

$$m_i(V_{ix} - v_{ix}) = \sum Q_{ix}, \qquad i = 1, 2, 3, 4 \tag{8.19}$$

and

$$m_i(V_{iy} - v_{iy}) = \sum Q_{iy}, \qquad i = 1, 2, 3, 4 \tag{8.20}$$

Summation is over the impulse components, indicated here by $Q_{ix}$ and $Q_{iy}$, acting on each body. For example, for body 1, $Q_{ix} = P_x$, $P_{1x}$, and $C_{1x}$. For angular motion,

$$I_i(\Omega_i - \omega_i) = \pm M + \sum r_{ix} Q_{ix} + \sum r_{iy} Q_{iy} \tag{8.21}$$

Summation here is of the moments of all impulse components; again $i = 1, 2, 3, 4$. The quantities $r_{ix}$ and $r_{iy}$ are moment arms to the impulse component relative to the mass center of each body. For example, for $Q_{1y} = P_y$, $r_{1y} = -d_1 \cos(\phi_1 + \theta_1)$. Four equations come from the conditions that the velocity in each body at the pin location must be the same as the corresponding point in the other body. These equations will have the form

$$V_{ij} + r_{ix}\Omega_i = V_{(i+2)j} + r_{(i+2)p}\Omega_{i+2}, \qquad i = 1, 2, \quad \text{and} \quad j = x, y \tag{8.22}$$

For example, the $x$ components of the final velocities at the pin of bodies 1 and 3 are such that

$$V_{1x} - d_{1p}\sin(\phi_{1p} + \theta_1)\Omega_1 = V_{3x} + d_{3p}\sin(\phi_{3p} + \theta_3)\Omega_3 \tag{8.23}$$

Three additional system equations come from the definition of the coefficients $e$, $e_m$, and $\mu$. These are, respectively,

$$(V_{1n} - V_{2n}) = -e(v_{1n} - v_{2n}) \tag{8.24}$$

$$e_m M = -(1 + e_m)\bar{I}(\Omega_2 - \Omega_1) \tag{8.25}$$

and

$$P_t = \mu P_n \tag{8.26}$$

In Eq. 8.24, for example,

$$V_{1n} = \left[V_{1x} - d_{1p}\sin(\phi_1 + \theta_1)\Omega_1\right]\cos\Gamma + \left[V_{1y} - d_{1p}\cos(\phi_1 + \theta_1)\Omega_1\right]\sin\Gamma \tag{8.27}$$

Equations 8.19–8.22 and 8.24–8.26 provide 19 system equations for which 19 unknown velocities and impulses can be obtained. If any external constraints exist, their equations and impulses must be added to the problem. Although they are not usually significant, it is possible to include moment impulses at pinned joints and at points of external constraints.

With no arbitrary constraints, the number of system equations is 19 and only numerical solution techniques are practical. Constraint equations can be added to the 19 or omitted as needed. This is a minor advantage of the Newtonian approach over the use of Lagrange's equations, where constraints are usually introduced using Lagrange multipliers or by the form of the kinetic energy expressions.

### 8.3.1 Rigid Barrier Impact of a Two-Member Body

One of the simplest, yet realistic, cases covered by the preceding system equations is the impact of a two-member body with a rigid barrier. This case will be discussed here in the context of an articulated road vehicle (tractor, semitrailer) colliding with a flat vertical barrier.[17] Local inertial axes $x$ and $y$ are used in the equations. Body 2, Fig. 8.7, is used to represent the barrier along the $t$ axis at an angle $\Gamma$. From the definition of restitution, the first equation is

$$\begin{aligned}\left[V_{1x} + d_1 \sin(\theta_1 + \phi_1)\Omega_1\right] \cos\Gamma + \left[V_{1y} - d_1 \cos(\theta_1 + \phi_1)\Omega_1\right] \sin\Gamma \\ = -e\Big\{\left[v_{1x} + d_1 \sin(\theta_1 + \phi_1)\omega_1\right] \cos\Gamma \\ + \left[v_{1y} - d_1 \cos(\theta_1 + \phi_1)\omega_1\right] \sin\Gamma\Big\} \qquad (8.28)\end{aligned}$$

From the definition of the impulse ratio,

$$P_y \cos\Gamma - P_x \sin\Gamma = \mu(P_y \sin\Gamma + P_x \cos\Gamma) \qquad (8.29)$$

The moment coefficient requires that

$$e_m M = (1 + e) I_1 \Omega_1 \qquad (8.30)$$

The common velocity constraint at the pinned connection between bodies 1 and 3 gives two equations; these are

$$V_{1x} - d_{1R} \sin(\theta_1 + \phi_{1R})\Omega_1 = V_{3x} + d_{3R} \sin(\theta_3 + \phi_{3R})\Omega_3 \qquad (8.31)$$

$$V_{1y} + d_{1R} \cos(\theta_1 + \phi_{1R})\Omega_1 = V_{3y} - d_{3R} \cos(\theta_3 + \phi_{3R})\Omega_3 \qquad (8.32)$$

The remaining six equations are those of linear and angular impulse and momentum for each of bodies 1 and 3. These are

$$m_1(V_{1x} - v_{1x}) = P_x + P_{1x} \tag{8.33}$$

$$m_1(V_{1y} - v_{1y}) = P_y + P_{1y} \tag{8.34}$$

$$I_1(\Omega_1 - \omega_1) = M + d_1 \sin(\theta_1 + \phi_1)P_x - d_1 \cos(\theta_1 + \phi_1)P_y - d_{1R} \sin(\theta_1 + \phi_{1R})P_{1x} + d_{1R} \cos(\theta_1 + \phi_{1R})P_{1y} \tag{8.35}$$

$$m_3(V_{3x} - v_{3x}) = -P_{1x} \tag{8.36}$$

$$m_3(V_{3y} - v_{3y}) = -P_{1y} \tag{8.37}$$

and

$$I_3(\Omega_3 - \omega_3) = -d_{3R} \sin(\theta_3 + \phi_{3R})P_{1x} + d_{3R} \cos(\theta_3 + \phi_{3R})P_{1y} \tag{8.38}$$

This set of system equations can be used to solve for the 11 unknowns: $V_{1x}$, $V_{1y}$, $\Omega_1$, $V_{3x}$, $V_{3y}$, $\Omega_3$, $P_x$, $P_y$, $P_{1x}$, $P_{1y}$, and $M$. Any constraints that may exist must be added and the preceding equations appropriately modified.

Let a velocity constraint be added to body 3 (trailer) such that the velocity of an arbitrary point ($d_{3c}$, $\phi_{3c}$, Fig. 8.7) must lie along a line with angle $\alpha_3$ measured counterclockwise from the $y$ axis. In that case,

$$\left[V_{3x} + d_{3c} \sin(\theta_3 + \phi_{3c})\Omega_3\right] \cos \alpha_3 + \left[V_{3y} - d_{3c} \cos(\theta_3 + \phi_{3c})\Omega_3\right] \sin \alpha_3 = 0 \tag{8.39}$$

Equations 8.36–8.38 must be modified to include the new unknown, impulse $C_3$. These become

$$m_3(V_{3x} - v_{3x}) = -P_{1x} + C_3 \cos \alpha_3 \tag{8.40}$$

$$m_3(V_{3y} - v_{3y}) = -P_{1y} + C_3 \sin \alpha_3 \tag{8.41}$$

and

$$I_3(\Omega_3 - \omega_3) = -d_{3R}\sin(\theta_3 + \phi_{3R})P_{1x} + d_{3R}\cos(\theta_3 + \phi_{3R})P_{1y}$$
$$+ d_{3c}\sin(\theta_3 + \phi_{3c})C_3\cos\alpha_3$$
$$- d_{3c}\cos(\theta_3 + \phi_{3c})C_3\sin\alpha_3 \qquad (8.42)$$

### 8.3.2 Application to Tractor, Semitrailer Barrier Impacts

Normally, all forces except the impact force are neglected during a collision. It is easy to show that for high-speed vehicle collisions, this is reasonable. A 3000 lb automobile that strikes a rigid barrier headon at 44 ft/s (30 mph), with a contact duration of 0.1 s, generates forces that exceed 100,000 lb. The wheel-to-ground friction forces are a fraction of the vehicle's weight (assuming a true, planar collision), are relatively small, and can be neglected. With articulated vehicles, the corresponding assumption may not be true. What is done here is to show with a hypothetical example how this assumption can be checked.

Table 8.4 lists the physical conditions corresponding to a typical unladen tractor, semitrailer truck. The tractor and semitrailer are in line and the tractor collides with a rigid vertical barrier at various angles $\Gamma$;

**TABLE 8.4 Physical Conditions for Rigid Barrier Impact**

| | | |
|---|---|---|
| Tractor mass/weight | 9500 kg | 20945 lb |
| Trailer rear wheel normal force | 36222 N | 8143 lb |
| Tractor yaw inertia | 21000 kg-m$^2$ | 15478 ft-lb-s$^2$ |
| Trailer mass/weight | 6000 kg | 13229 lb |
| Trailer yaw inertia | 48000 kg-m$^2$ | 35375 ft-lb-s$^2$ |
| Initial speed | 13.41 m/s | 30 mph |
| Initial system energy | $2.65 \times 10^6$ N-m | $1.95 \times 10^6$ ft-lb |
| Orientation, $\theta_1 = \theta_3$ | 0° | |
| Coefficient of restitution | 0.05 | |
| Moment coefficient | −1.0 (no moment) | |
| Vehicle/barrier friction coefficient, $f$ | 0.2 | |

a value of $\Gamma = 0$ represents a head on collision. Frictional impulses of all tractor wheels are neglected. The condition is taken where all wheels are freely rolling, that is, no braking. A constraint is placed centrally between the trailer wheels and constrains the velocity to lie along the heading angle of the trailer. As a result, $\alpha_3 = 90°$ for $\theta_1 = \theta_3 = 0°$. This simulates a transverse frictional wheel-to-ground impulse from the rear trailer wheels large enough to maintain the condition of rolling. If this impulse exceeds the capacity of friction, then it is concluded that the rear of the trailer can attain a transverse velocity due to the impact.

For $\Gamma = 0°$, the collision is directly headon and neither tractor nor trailer will generate a final angular velocity. Initial and final velocities automatically lie along the constraint direction and so the constraint impulse is zero. For small barrier angles, $\Gamma$, the tractor is not sliding at separation and develops a positive final angular velocity. As $\Gamma$ is increased, $\mu_0$ is reached and the impact becomes a glancing one for which the final tractor angular velocity is negative. The angular velocity of the tractor through the pin connection transmits or causes a corresponding angular velocity of the trailer, which also changes sign. This in turn causes the constraint between the rear wheels to change sign as seen in Fig. 8.8. At $\Gamma = 90°$, a collision does not occur and the constraint impulse goes to zero. The maximum constraint impulse occurs for barrier angles between 30° and 45° for the physical values of this example. The peak value of the constraint impulse is 22,500 N-s.

For the ideal conditions of a triangular force shape and a contact duration of 0.15 s, the peak transverse constraint force at the semitrailer wheels is 300,000 N. No realistic value of wheel-to-ground friction coefficients could supply a force large enough to prevent the rear trailer wheels from sliding due to the impact. For barrier angles between 15° and 20°, the converse can be true however. Other conditions could be investigated, such as breakout of the rear tractor wheels, different weights and weight distributions of the trailer, and so on.

When the constraint impulse is large enough to overcome friction, the velocity constraint can be removed and the problem solved anew, permitting a transverse velocity change at the trailer rear wheels. The new solution can neglect the transverse frictional impulse at the trailer wheels or include it as a known external impulse, whichever is appropriate. If an applied, or known, external impulse is desired, it can be handled by dropping the velocity constraint, Eq. 8.39, and treating $C_3$ in Eqs. 8.40–8.42 as known.

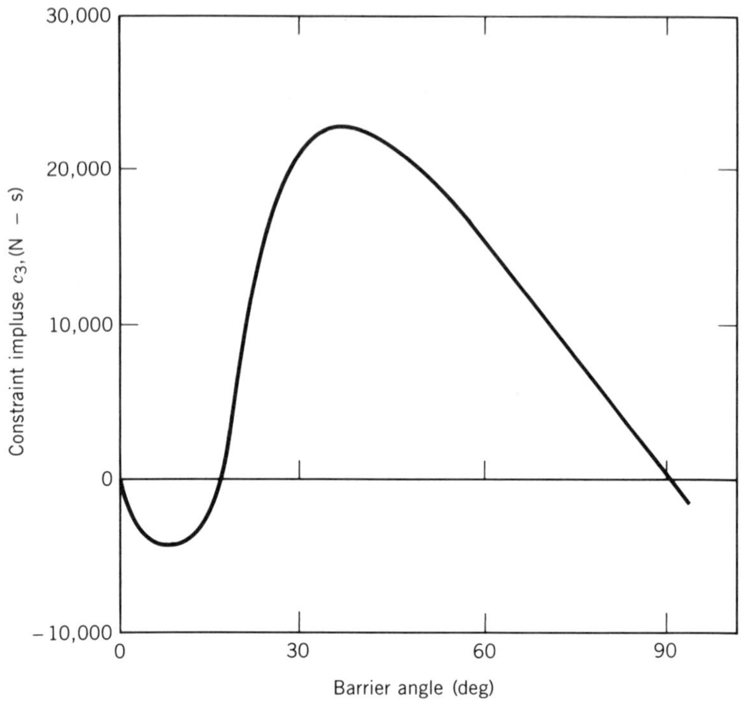


**Figure 8.8.** Constraint impulse in a two-member chain impact with a rigid surface; $f = 0.2$. (Zero degrees is a perpendicular impact.)

## 8.4 PROBLEMS

**1.** Solve Example 8.1 for the condition that mass 3 is grasped but remains free to rotate. That is, find $\Omega_3$ for $e_m = -1$.

**2.** Suppose $m_3$ in Fig. 8.2 forms a rigid barrier at angle $\alpha$, and $m_1$ and $m_2$ are configured such that $\phi_1 = \phi_2$ (see Section 8.1.2). For $e_m = -1$, devise a scheme for solving the system equations. (Note that the right-hand sides of Eqs. 8.4 and 8.5 are identical and become singular.)

## REFERENCES

1. Shabana, A. A., *Dynamics of Multibody Systems*, John Wiley, New York, 1989.

2. Kamman, J. W. and Huston, R. L., "Dynamics of Constrained Multibody Systems," *Trans. ASME, J. Appl. Mech.*, 84-WA/APM-25.
3. Pennock, G. R. and Yang, A. T., "Dynamic Analysis of a Multi-Rigid-Body Open-Chain System," *Trans. ASME, J. Mech., Transmissions and Automation in Design*, **105,** 28–34, 1983.
4. Greenwood, D. T., *Principles of Dynamics*, Prentice-Hall, Englewood Cliffs, NJ, 1988.
5. Lanczos, C., *The Variational Principles of Mechanics*, University of Toronto Press, 1962.
6. Goldsmith, W., *Impact*, Edw. Arnold, London 1960.
7. Huang, T. C., *Engineering Mechanics, Vol. 1, Statics*, Addison Wesley, Reading, MA, 1967.
8. Meriam, J. L., *Engineering Mechanics, Statics and Dynamics*, John Wiley, New York, 1978.
9. Kane, T. R. and Levinson, D. A., *Dynamics: Theory and Applications*, McGraw-Hill, New York, 1985.
10. Tseng, C-S., Crane, C., and Duffy, J., "Generating Collision-Free Paths for Robot Manipulators," *Computers in Mechanical Engineering*, Sept./Oct., pp. 58–64, 1988.
11. Eppinger, S. D. and Seering, W. P., "On Dynamic Models of Robot Force Control," Proceedings of the International Conference on Robotics and Automation, IEEE, New York, 1986.
12. Khatib, O. and Burdick, J., "Motion and Force Control of Robot Manipulators," Proceedings of the International Conference on Robotics and Automation, IEEE, New York, 1986.
13. Salisbury, J. K. and Roth, B., "Kinematic and Force Analysis of Articulated Mechanical Hands," *J. Mechanisms, Transmissions and Automation in Design*, **105,** 35–41, 1983.
14. Nye, T. W., LeBlanc, D. J. and Cipra, R. J., "Design and Modeling of a Computer-Controlled Planar Manipulator," *Int. J. Robotics Res.*, **6**(1), 85–95, 1987.
15. LeBlanc, D. J., "Modeling, Trajectory Calculation and Control of a Planar Manipulator," Master's Thesis, Purdue University, Lafayette, IN 1985.
16. Ferrara, S. A., "Impact of a Planar Manipulator," Master's Thesis, University of Notre Dame, 1989.
17. Brach, R. M., "Impact of Articulated Vehicles," SAE Technical Paper 860015, 1986.

# CHAPTER 9

# VIBRATORY IMPACT

In all other chapters of this book, impacts are viewed as single events. A broad area of applications exists where these events are repetitive. This is referred to as vibratory impact or simply vibroimpact and is covered in this chapter. A common and most useful type of vibratory impact is steady-state impact, which is a periodic motion for which the effects at the initial or starting conditions are no longer evident or significant. There are many similarities to and some important differences from the well-known steady-state vibration problem where an external source of energy (harmonic input) and an energy dissipation mechanism (damping) reach a state of balance such that the system's motion becomes repetitive or cyclic with a constant amplitude. In the case of vibratory impact, some or all of the energy is dissipated through the collision process, which is regulated by the coefficients $e$, $e_m$, and $\mu$. Since energy is lost during each impact, an external energy supply must exist in order to maintain steady vibratory motion. Input forces may be external and harmonic as in vibration problems, but may also be part of the impact process itself, such as a tennis player idly bouncing a ball up and down with a racket. Other possibilities exist for the input as well. A major difference exists between vibratory impact and the well-known steady-state vibration problem and is centered on the concept of linearity. The steady-state vibration problem with its harmonic force and a linear mass–

spring–damper system[1,2] leads to a linear second-order differential equation and harmonic motion. The classical view of impact involves an instantaneous velocity change. When repetitive impacts occur as part of vibratory motion of a linear system, the problem becomes nonlinear. In fact, in the context of repetitive motion, it appears that the impact process is always nonlinear. This may be surprising since the system equations for the impact process are linear and algebraic and the equation of motion for the physical system between impacts may be linear. However, the coupling of a discontinuous motion (impact) to the motion of a linear second-order system results in a nonlinear system. In other words, the coupling is responsible for the nonlinearity.

The existence of nonlinearities in dynamical systems can drastically increase the complexity of the problems. For example, with a linear system, the frequency of the output motion is identical to the input frequency; this is not so for nonlinear systems. Discovery of the frequency content of the output is one of the central problems of nonlinear dynamic analysis. One of the problems covered here is a ball or particle bouncing from a horizontal surface, which itself oscillates vertically at a fixed rate or frequency. The ball is shown to be capable of exhibiting periodic motion at the table's frequency, but under certain conditions it can have a nonperiodic, or chaotic, motion.[3,4] Chaotic motion has the appearance of randomness, but, in fact, it is quite deterministic.

Three basic types of vibratory impact are covered here, to a large extent by example. All involve a single point mass or particle, that is, no rigid body rotation. The first is an example of periodic, steady-state motion of a particle bouncing between rigid attractive surfaces. The second is the bouncing ball problem whose solution exhibits chaotic motion. The last type of problem involves mass–spring–damper systems that undergo impacts. The topic of phase-space analysis is found to be very useful in nonlinear dynamics. This is discussed and illustrated with a vibratory impact problem.

## 9.1 PERIODIC POINT-MASS IMPACTS

For a point mass to undergo vibratory motion, some fundamental properties of motion must exist. To be oscillatory or vibratory, reversals in direction must take place. With the exception of the ideal case of frictionless and perfectly elastic impact, energy is lost in the impact process and so an external source of energy must exist to provide sustained oscillations. Colver[5] documents an excellent example of such a problem, which involves aerosol particles. In this problem a particle is initially at rest on a flat horizontal plate located below a second parallel plate a distance $d$ above. The plates are given an applied dc electric charge causing them to form a large parallel-plate capacitor. While in physical contact with the lower plate, the particle receives a like charge and is immediately attracted toward the oppositely charged, upper plate. If the charge and field are sufficient to induce an attractive acçeleration greater than $g$, the local gravitational attraction, the particle will rise, gain velocity, and impact the upper plate. When it contacts the upper plate, the particle's polarity reverses and is then attracted by the lower plate. The entire process then continues with the particle bouncing up and down in a vibratory manner. If the plates are oriented vertically, a tangential velocity develops due to gravity or if the plates are not parallel and diverge somewhat, the particle "flows" while alternately impacting the two plates. This is used in a practical device as a means of generating aerosol streams.[6]

Figure 9.1 illustrates the $i$th impact pair of a particle bouncing between charged plates spread at a small included angle of $2\beta$ about a horizontal plane. The position $x_{i+1}$ and the velocity $V_{i+1}$ are sought based on the parameters of the system under the following assumptions:

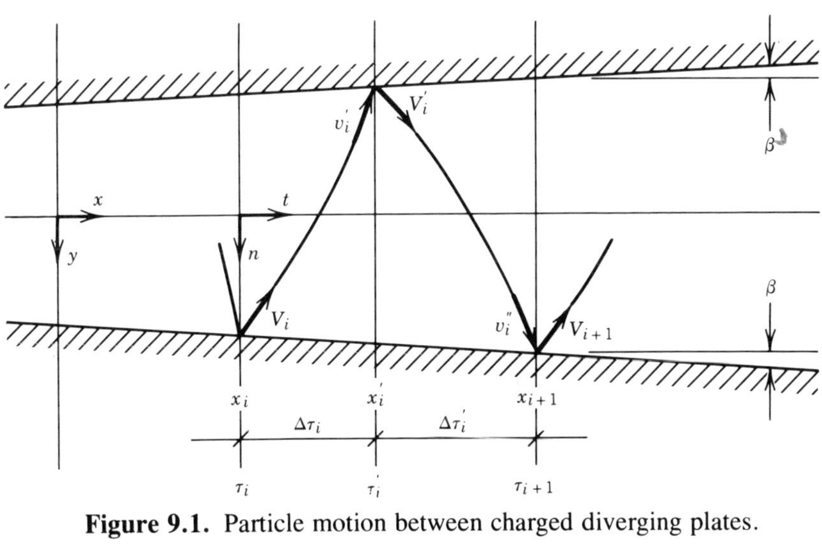


**Figure 9.1.** Particle motion between charged diverging plates.

1. motion is in a plane perpendicular to the charged plates,
2. the angle $\beta$ is small,
3. no aerodynamic drag,
4. coefficient of restitution is a constant,
5. a uniform electrostatic field exists perpendicular to the centerline between plates.

The presence of gravity makes the problem vertically unsymmetrical. As a result, a cycle is taken to include a pair of consecutive impacts, one each from the upper and lower surfaces. Based on the particle impact model of Chapters 2 and 3, the velocity components following the $i$th impact at the upper surface are

$$V'_{n_i} = -ev'_{n_i} - \left(e + 1 - \frac{\mu}{\mu'_c}\right)\beta v'_{t_i} \tag{9.1}$$

$$V'_{t_i} = -\left(e + 1 - \frac{\mu}{\mu'_c}\right)\beta v'_{n_i} + \left(1 - \frac{\mu}{\mu'_c}\right)v'_{t_i} \tag{9.2}$$

where primed quantities correspond to the upper surface impact and where the small-angle approximation allows $\beta$ to be used for sin $\beta$ and 1 for cos $\beta$. The quantity $\mu'_c$ is the critical impulse ratio $(v'_{t_i}/v'_{n_i})/(1 + e)$ from Eq. 3.13. For the second impact,

$$V_{n(i+1)} = -ev''_{n_i} + \left(e + 1 - \frac{\mu}{\mu''_c}\right)\beta v''_{t_i} \tag{9.3}$$

$$V_{t(i+1)} = \left(e + 1 - \frac{\mu}{\mu''_c}\right)\beta v''_{n_i} + \left(1 - \frac{\mu}{\mu''_c}\right)v''_{t_i} \tag{9.4}$$

It is important to note that the critical value of $\mu$ changes from collision to collision. For nearly horizontal plates (small $\beta$) the upward and downward accelerations of the particle differ because of gravity. They are

$$a_u = (\gamma - 1)g \tag{9.5}$$

and

$$a_d = (\gamma + 1)g \tag{9.6}$$

where $g$ is the gravitation constant and $\gamma$ is the ratio of the electrostatic force to the particle's weight. The ratio $\gamma$ must be greater than 1 for the particle to rise from the lower to the upper plate. With uniform acceleration between impacts,

$$v'_{n_i} = V_{n_i} - (\gamma - 1)g\Delta\tau_i \tag{9.7}$$

and

$$v'_{t_i} = V_{t_i} \tag{9.8}$$

The position of the particle at the time of impact with the upper surface is

$$y(\tau'_i) = y(\tau_i + \Delta\tau_i)$$

or

$$-\frac{d}{2} - \beta x'_i = \frac{d}{2} + \beta x_i + V_{n_i}\Delta\tau_i - \tfrac{1}{2}(\gamma - 1)g\Delta\tau_i^2 \tag{9.9}$$

and

$$x(\tau'_i) = x(\tau_i + \Delta\tau_i)$$

or

$$x'_i = x_i + V_{t_i}\Delta\tau_i \tag{9.10}$$

Equations 9.9 and 9.10 are two equations solvable for $x'_i$ and $\Delta\tau_i$, where

$$\Delta\tau_i = \frac{1}{(\gamma - 1)g}\Big\{V_{ni} + \beta V_{ti} + \big[(V_{n_i} + \beta V_{ti})^2 + 2(\gamma - 1)g(d + 2\beta x_i)\big]^{1/2}\Big\} \tag{9.11}$$

with $x_i'$ found from Eq. 9.10. The downward motion has the velocity components

$$v_{n_i}'' = V_{n_i}' + (\gamma + 1)g\Delta\tau_i' \tag{9.12}$$

and

$$v_{ti}'' = V_{t_i}' \tag{9.13}$$

It impacts that lower surface at

$$y(\tau_{i+1}) = y(\tau_i' + \Delta\tau_i')$$

$$\frac{d}{2} + \beta x_{i+1} = -\left(\frac{d}{2} + \beta x_i'\right) + V_{n_i}'\Delta\tau_i' + \tfrac{1}{2}(\gamma + 1)g\Delta\tau_i'^2 \tag{9.14}$$

and

$$x(\tau_{i+1}) = x(\tau_i' + \Delta\tau_i')$$

or

$$x_{i+1} = x_i' + V_{t_i}'\Delta\tau_i' \tag{9.15}$$

Solution for $\Delta\tau_i'$ gives

$$\Delta\tau_i' = \frac{1}{(\gamma + 1)g}\Big\{-(V_{n_i}' - \beta V_{t_i}') + \big[(V_{n_i}' - \beta V_{t_i}')^2 + 2(\gamma + 1)g(d + 2\beta x_i')\big]^{1/2}\Big\} \tag{9.16}$$

with $x_{i+1}$ obtained from Eq. 9.15. These equations are solved sequentially for positions and velocities to give the particle's motion as it bounces from surface to surface.

These equations simplify considerably for the condition of parallel plates ($\beta = 0$) and frictionless impacts. If the particle has no initial tangential velocity, it simply bounces up and down with $x = 0$. This is the case studied in Ref. 5. If it has an initial tangential velocity, the frictionless condition implies that it moves horizontally with a constant velocity as it bounces between the plates. The final normal velocity from the lower surface following the $i$th impact pair is

$$V_{n(i+1)}^2 = 2e^2dg\left[1 + \gamma + e^2(\gamma - 1) + e^2\, V_{n_i}^2/2dg\right] \quad (9.17)$$

After the $(i + 1)$th impact pair

$$V_{n(i+2)}^2 = 2e^2dg\left\{\left[1 + \gamma + e^2(\gamma - 1)\right]\left[1 + e^4\right] + e^6\, V_{n_i}^2/dg\right\} \quad (9.18)$$

after the $(i + N)$th impact pair

$$V_{n(i+N)}^2 = 2e^2dg\left\{\left[\gamma + 1 + e^2(\gamma - 1)\right]\left[1 + e^4 + e^8 + e^{10} + \cdots + e^{4N}\right] + e^{4N+2}v_{n_i}^2/2dg\right\} \quad (9.19)$$

Let $i = 0$ and assume the particle starts from rest; $V_{n0} = 0$. Then

$$V_{nN}^2 = 2e^2dg\left\{\left[\gamma + 1 + e^2(\gamma - 1)\right]\sum_{j=0}^{N} e^{4(j)}\right\} \quad (9.20)$$

If the number of collisions is allowed to become unbounded, the series in Eq. 9.20 converges to the finite sum of $(1 - e^4)^{-1}$. This gives a steady-state velocity of

$$V_{ns}^2 = \frac{2e^2dg}{1 - e^4}\left[\gamma + 1 + e^2(\gamma - 1)\right] \quad (9.21)$$

where $V_{ns}$ is the velocity of rebound from the lower surface. The presence of gravity makes the rebound from the upper surface differ from $V_{ns}$ and, in the preceding discussion, can be found to be

$$V_{ns}'^2 = \frac{2e^2dg}{1 - e^4}\left[\gamma - 1 + e^2(\gamma - 1)\right] \quad (9.22)$$

With these, any related steady-state quantity of interest can be found, such as the average velocity over a cycle or the steady-state period. The latter is

$$\Delta\tau = \left[\frac{2d}{g(1 - e^4)}\right]^{1/2} \cdot \left\{\frac{(\gamma + 1) - e(\gamma - 1)}{(\gamma - 1)(\gamma + 1)}\left[\gamma + 1 + e^2(\gamma - 1)\right]^{1/2} + \frac{(\gamma - 1) - e(\gamma + 1)}{(\gamma - 1)(\gamma + 1)}\left[\gamma - 1 + e^2(\gamma + 1)\right]^{1/2}\right\} \qquad (9.23)$$

If the particle's weight is small compared to the electrostatic force, that is, $\gamma >> 1$, then the steady-state velocity is

$$V_{ns} = \left(\frac{2e^2 d\gamma g}{1 - e^2}\right)^{1/2} \qquad (9.24)$$

The corresponding period is

$$\Delta\tau = \left(\frac{8d}{\gamma g}\frac{1 - e}{1 + e}\right)^{1/2} \qquad (9.25)$$

with the steady-state frequency of $f = 1/\Delta\tau$. The velocity and period are proportional to the square root of the plate separation distance. (If the particle's diameter is not small compared to the plate separation distance, then $d - 2r$ should be used for $d$ in the preceding equations, where $r$ is the radius of the particle.) The steady-state vibratory motion for frictionless parallel plates can be shown to be asymptotically stable for $e < 1$ (see Problem 9.2). This means it will always reach the steady-state velocities given previously regardless of the initial velocity and any small perturbations to the motion. With the particle initially at rest, the time at which steady-state velocity is reached depends among other things on the coefficient $e$. This is illustrated in Fig. 9.2, which shows the rebound velocity $V_{ns}$ as a function of the number of collisions as given by Eq. 9.20 for plates a distance of 0.05 m apart. Not only does the time to reach steady-state velocity depend on $e$, but the velocity itself does also as given by Eq. 9.24.

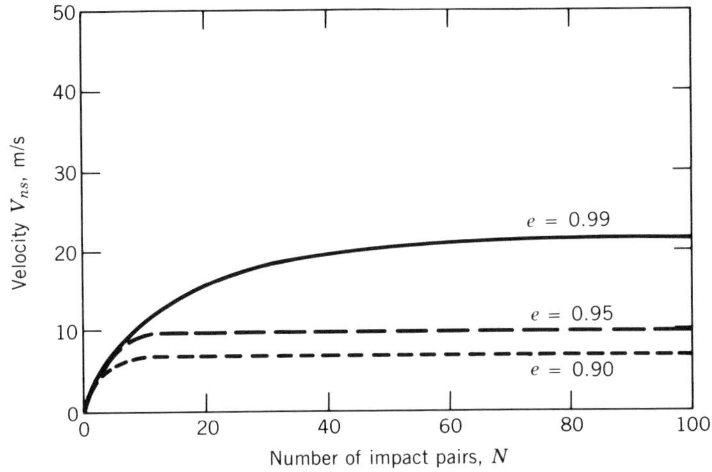


**Figure 9.2.** Rebound velocity growth, charged parallel plates 0.05 m apart; no friction and $\gamma = 10$.

## 9.2 PHASE PLANE

Because vibratory impact problems are nonlinear, methods from the field of nonlinear dynamics are often used. One of the most useful techniques from this field is graphical in nature and displays the motion of the system in a *phase plane.*[7] The phase plane is a plot of velocity versus position and is also sometimes referred to as state space. Some of the simpler features of phase plane analysis are covered here.

Consider an ideal oscillator consisting of a mass $m$ connected to a linear spring with rate $k$ and with no impacts. From vibration theory,[1] the motion is known to be harmonic and conservative, that is, periodic with no energy lost. The total energy is given by

$$E_0 = \tfrac{1}{2}mv^2 + \tfrac{1}{2}kx^2 \tag{9.26}$$

where $v = \dot{x}(\tau)$ is velocity, $x = x(\tau)$ is position. Equation 9.26 can be put into a more convenient form by dividing by $k/2$, which gives

$$\frac{v^2}{\omega_n^2} + x^2 = E_0' \tag{9.27}$$

The quantity $\omega_n$ is the natural frequency and $E_0' = 2E_0/k$. Since the energy is a constant, $E_0$ must equal the total energy in the system initially (at $\tau = 0$) and depend on the initial conditions. Equation 9.27 is an equation of an ellipse in the $v$, $x$ plane with initial energy as a parameter. Some typical curves are shown in Fig. 9.3 for three different values of $E_0$.

Each point of a phase plane curve such as $P$ represents a state of the system. As time progresses, the state changes and the point sweeps out a curve in phase space called a trajectory. The elliptical trajectories in Fig. 9.3 for the ideal linear oscillator are closed and indicate periodic motion. If the position of the moving point $P$ is projected vertically onto a time axis, the position as a function of time results. Horizontal projection parallel to the $x$ axis results in the velocity.

The motion associated with a set of initial conditions in phase space describes the time-dependent behavior of a dynamical system. In a *conservative* autonomous system, the area of a closed curve in two-dimen-

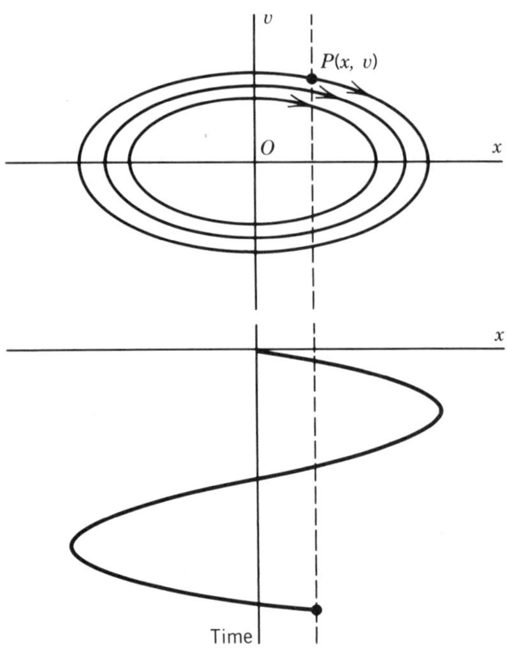


**Figure 9.3.** Sample phase plane curves (trajectories) and displacement–time response for a linear oscillator.

sional phase space (or volume of a surface in three-dimensional space) representing initial conditions does not shrink as time goes. In a *dissipative* system, however, such as the ones discussed here, the motion following from initial conditions in a given neighborhood asymptotically tend toward a set called an *attractor*. The simplest possible attractor is a point, indicating that the corresponding long-term behavior of such a system is *static equilibrium*. Another possible attractor is a closed curve; this represents periodicity and the closed curve itself is called a *limit cycle*. In a similar fashion, one can have a toroidal attractor in a phase space of dimension three or greater. If the ratios of the two frequencies which compose motion on the surface of a three-dimensional torus is rational, the system repeats itself in a *periodic* manner; if the ratio is irrational, the system does not repeat itself exactly and is *quasiperiodic*. It is also possible for the behavior to be completely aperiodic, in which case the attractor is referred to as *strange* or *chaotic*. If a single attractor exists to which all initial conditions are attracted, it is said to be globally stable. However, a given dynamical system may, for the same set of parameters, have more than one attractor. In this case the *basin of attraction* of each attractor determines the set of initial conditions that will end up in the vicinity of that attractor.

### 9.2.1 Trajectories of the Charged Plate Problem

The phase plane curves for vibratory impact of a particle oscillating between parallel charged plates is discussed here. Between plates, the acceleration is a constant; see Eqs. 9.5 and 9.6. When rising toward the upper plate, the velocity is

$$v_n = -\left[V_i^2 - 2g(\gamma - 1)(y - d/2)\right]^{1/2} \qquad (9.28)$$

and when moving toward the lower plate it is

$$v_n = \left[V_i'^2 + 2g(\gamma + 1)(y + d/2)\right]^{1/2} \qquad (9.29)$$

When it reaches the upper plate, the normal velocity change is such that $V_i' = -ev_i'$ and at the lower plate, $V_{i+1} = -ev_i''$. Figure 9.4 shows the phase plane trajectory of a particle starting from rest on the lower plate, which is represented by point $B_0$. The particle accelerates toward the upper plate given by the curve $B_0A_1$. After a velocity change, rebound

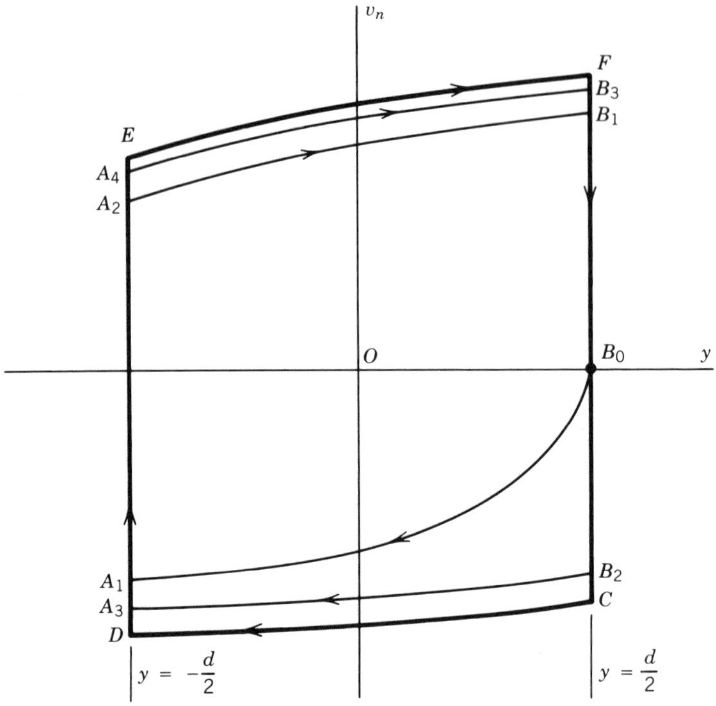


**Figure 9.4.** Phase plane trajectories and limit cycle for a particle bouncing between charged plates.

occurs at point $A_2$ with acceleration toward the lower plate along the trajectory $A_2B_1$. Rebound from the lower plate is at point $B_2$ and trajectory $B_2A_3$ follows. This continues until steady state is reached, which is represented by the closed curve *CDEF*, which is a limit cycle for this motion. The steady-state velocities $V_{ns}$ and $V'_{ns}$ given by Eqs. 9.21 and 9.22 are the velocities at points *C* and *E*, respectively. Note that the limit cycle is not symmetric with respect to the displacement axis due to the bias of gravity. Because this limit cycle is globally stable, all trajectories approach it whether or not they start from $B_0$.

## 9.2.2 The Doorbell Problem

Another illustration of the use of the phase plane for vibratory impact is presented. It is an analysis of an idealized, old-fashioned doorbell shown schematically in Fig. 9.5 and follows from the work of Akashi.[8] When

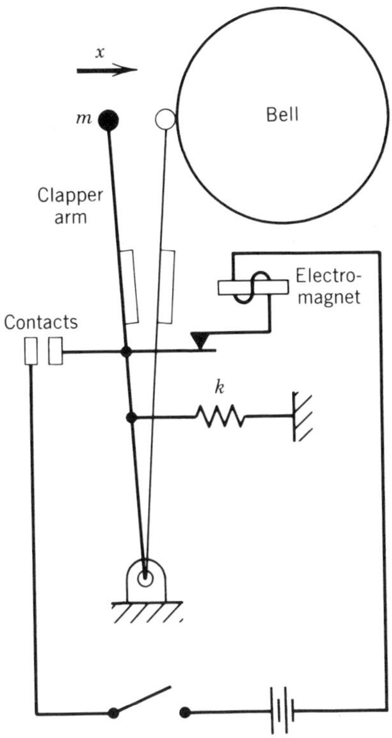


**Figure 9.5.** Schematic diagram of a doorbell.

the switch is closed, a constant electromagnetic force $F$ attracts the clapper mass $m$ toward the circular bell. When the mass reaches a position $x = x_0$, the contacts open, the force is cut off and the mass continues on to impact the bell at $x = x_b$. The mass rebounds with coefficient $e$ and the magnetic force returns when the contacts close at $x = x_c$. It continues the rebound until the spring and force reverse the clapper and it proceeds toward the next impact. If the equivalent mass is $m$ and the equivalent stiffness is $k$, then the trajectories lie on segments of ellipses such as in Fig. 9.3 between impacts. When the force is zero, the energy of the system is

$$\tfrac{1}{2}m\dot{x}^2 + \tfrac{1}{2}kx^2 = E_0 \tag{9.30}$$

During the time the contacts are closed, the force is applied and the oscillation takes place about a position biased by the force such that

$$\tfrac{1}{2}m\dot{x}^2 + \tfrac{1}{2}k(x - F/k)^2 = E \tag{9.31}$$

In this case, the trajectories remain elliptical in shape but are centered at $x = F/k$. These variables can be nondimensionalized by letting $u = kx/F$. If the time variable $\tau$ is replaced by $t = \omega_n \tau$, then the velocity $\dot{u} = \omega_n \, du/d\tau$, where $\dot{u} = du/dt$ and $\omega_n = \sqrt{k/m}$. Then Eqs. 9.30 and 9.31 become

$$\dot{u}^2 + u^2 = \text{const} \tag{9.32}$$

when the contacts are open and

$$\dot{u}^2 + (u - 1)^2 = \text{const} \tag{9.33}$$

when closed. The phase plane curves are now circular arcs and the trajectories alternate between circles centered at 0 and 1 as shown in Fig. 9.6. Each cycle includes an impact when $u = u_b = kx_b/F$. The trajectories switch circles at the positions at which the contacts open and close, $u_0 = kx_0/F$ and $u_c = kx_c/F$, respectively. Akashi shows that for a properly selected set of parameters the trajectories approach a limit cycle with an equal radius $r$ for both circular portions of the trajectories, where

$$r^2 = (1 + u_b^2 - 2u_0) + 2(u_0 - u_c)/(1 - e^2) \tag{9.34}$$

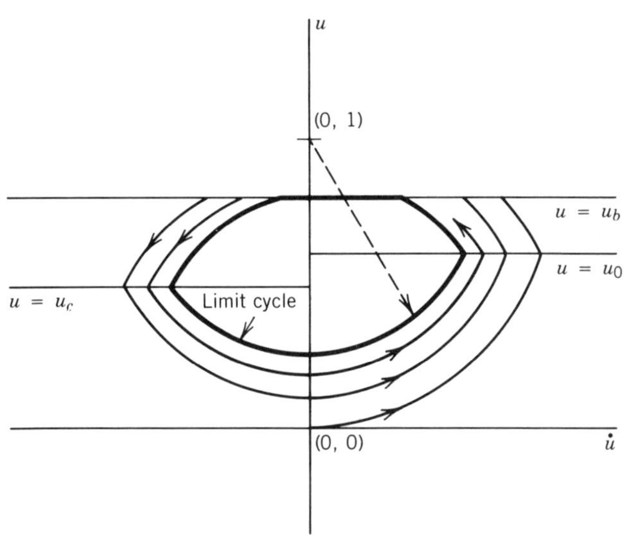


**Figure 9.6.** Doorbell: phase plane trajectories and limit cycle.

Note that each impact is characterized by a straight line segment of the trajectories at $u = u_b$, the position of the bell. Figure 9.6 shows the trajectories and limit cycle for a ringing bell. The doorbell is capable of other types of motion, depending on the choices of the system parameters such as $F$, $k$, $m$, $u_0$, $u_c$, and $u_b$. Akashi shows, for example, that the bell can be made to ring vigorously, slowly die out, and become silent. Other more erratic or chaotic motion is even possible. These types of motion are illustrated and described in the next sections for other systems. This and prior sections show that the phase plane can be used conveniently to illustrate vibratory motion with impact.

## 9.3 POINT-MASS VIBRATORY IMPACT: CHAOTIC MOTION

A new example problem is now analyzed in this section in order to illustrate some different and exotic features of vibratory impact. The problem consists of a ball continually bouncing above a massive table while the table translates in a circular path in a vertical plane with a given frequency. The table moves simultaneously up and down and side to side and the ball is assumed to move in a plane of motion of the table. The ball is modeled as a point mass. Since rotation is not considered, the impact model is that of a single particle and follows from the equations of Chapter 2.

This problem has some similarities and many differences from those already discussed. Consider the particle bouncing between the charged plates. In that problem, the impact model consists of a particle rebounding from a stationary surface. Here the surface is moving. The steady frequency of motion between the charged plates is controlled by the parameters such as field strength and separation distance. Here, the table frequency contributes to timing of the motion, but so does the time of flight of the ball's trajectory. A greater variability in the potential motion of the bouncing ball might be expected. This aspect of motion is closer to the doorbell problem. In that case, no fixed frequency exists for the opening and closing of contacts; it depends on the clapper's motion. The bell problem and bouncing ball problem have another similarity: only one impact surface exists. The types of restoring or returning forces differ, however. In the case of the doorbell, the applied force and spring return the clapper, whereas gravity returns the ball for the next bounce.

The motions of the bell and bouncing ball possess a common trait that the parallel plate problem does not. This is related to the concept of stability. Whenever a system is nonlinear, the motion can display different characteristics as parameters are changed. Once the particle between the plates begins to move, it always reaches steady state; it can be shown that its steady-state motion will only be changed temporarily by a small disturbance (the presence of a surface imperfection, for example). The trajectory returns to the limit cycle in the phase plane; see Problem 9.2. This is considered to be a characteristic of stable motion. Under certain conditions (not discussed earlier) trajectories of systems such as the doorbell and the bouncing ball (as will be seen) can change drastically when system parameters or initial conditions are perturbed by small amounts. This extreme sensitivity to initial conditions is one characteristic of what is called chaotic motion.

Another essential difference between the parallel plate problem and the bouncing ball problem pertains to the number of impacts per cycle. For steady-state motion, the particle alternately bounces from the lower to upper surface and back. It never has multiple bounces from one surface in a cycle. As mentioned, limit cycles exist other than the one presented for the doorbell. A large variety of limit cycles are found to exist for the bouncing ball. Many of them have multiple impacts per cycle, where the concept of cycle is based on the observation of repetitive motion and is not necessarily determined by the time of each bounce or the timing of the table.

Perhaps the most significant difference between the character of motion of the bouncing ball and the parallel plate problems is that the ball can possess a randomlike motion for some frequencies of table motion. Although random in appearance, the motion is found to have a definite underlying structure. This type of motion is called *chaos* and is the subject of intense research now and in the recent past.[4,9–15,18] This type of motion will be encountered in the following discussion of the bouncing ball problem.

### 9.3.1 Equations (Nonlinear Mapping)

Figure 9.7 illustrates the geometry and coordinates of a ball $m_1$ bouncing from the surface of a massive table $m_2$. The surface translates with a circular motion with velocity components given by

$$V_{2n} = A\omega \cos(\omega\tau) \tag{9.35}$$

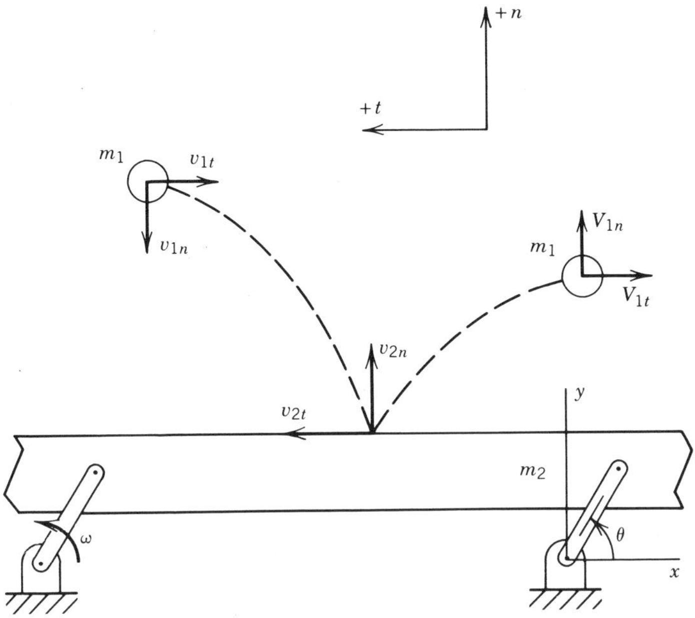


**Figure 9.7.** Particle bouncing over an oscillating table.

and

$$V_{2t} = A\omega \sin(\omega\tau) \tag{9.36}$$

where $A$ is the amplitude, $\omega$ is the circular frequency and $\tau$ is time. Between impacts, the ball has a parabolic trajectory above the surface. The $(k + 1)$st impact occurs when the ball and surface reach the same vertical position for the first time following the $k$th impact. The vertical position of the surface is found from integration of Eq. 9.35. The vertical position of the ball is found from the parabolic equations of uniform acceleration trajectories. Equating them gives an equation that can be solved to furnish the time interval between the $k$th and $(k + 1)$st impact, that is,

$$A \sin \omega(\tau_k + t_k) - V_{1n}^k t_k - g t_k^2/2 - A \sin \omega\tau_k = 0 \tag{9.37}$$

where the time interval $t_k = \tau_{k+1} - \tau_k$. The initial normal and tangential components of the velocity at the $(k + 1)$st impact are found from the final velocity components of the $k$th impact, which are used as the initial conditions of the trajectory equations. These velocities and Eqs. 2.12 and 2.14 (for $m_1/m_2 = 0$) give the final velocity components of the $(k + 1)$st impact. These are

$$V_{1n}^{k+1} = -e(V_{1n}^{k} - gt_k) + (1 + e)A\omega \cos \omega\tau_{k+1} \qquad (9.38)$$

and

$$V_{1t}^{k+1} = V_{1t}^{k} + \mu(1 + e)(A\omega \cos \omega\tau_{k+1} - V_{1n}^{k} - gt_k) \qquad (9.39)$$

The last three equations provide a recursive solution to the bouncing ball problem. Equation 9.37 is solved for $t_k$. Equations 9.38 and 9.39 then provide the velocity components following the $(k + 1)$st impact. This is repeated as often as necessary. Of course $\mu$ in Eq. 9.39 must be properly chosen as $f$ or $\mu_c$ for each impact. Because Eq. 9.37 is transcendental, its solution for $t_k$ must be found numerically. All of the results to follow are generated numerically.

In the parlance of nonlinear dynamics, the preceding three solution equations are called a (discrete) mapping since the equations "map" the $k$th bounce onto the $(k + 1)$st. Since the system is nonlinear, this is a nonlinear mapping. Four parameters exist in this problem. These are

- coefficient of restitution, $e$
- coefficient of friction, $f$
- surface amplitude, $A$
- surface frequency, $\omega$

The solution also depends on the starting or initial values of ball velocity components.

### 9.3.2 Motion

Two broad categories of long-term motion result from solution of the mapping equations; these are periodic and nonperiodic. The periodic motions imply cyclic or repetitive motion. In general for linear dynamical systems, motions have the same frequency as the input (here the table frequency).[7] With nonlinear systems, this is not always the case. For one impact per cycle, with each trajectory identical to the next and the same table position and velocity at the time of impact, the motion is obviously periodic. Other types of periodic motion exist however. For example, two trajectories can succeed each other, with the pairs of impacts and trajectories then repeating. A repetitive single impact and trajectory will be called period-one motion. Repetitive pairs of impacts and trajectories are called period-two with the general term being "period-$N$." Here there are $N$ impacts and trajectories continually followed by the same $N$ impacts and trajectories.

For example, let $e = 0.9, f = 0.4$, and $A = 0.015$ m. Starting from various initial conditions and after hundreds of table oscillations when the table frequency is 25.5 rad/s, period-one motion emerges as a steady motion. If the table frequency is increased to $\omega = 26.0$ rad/s, then a steady period-two motion emerges. Figure 9.8 shows phase plane diagrams of these period-one and period-two motions. They are closed curves and are stable limit cycles, since any small, transient change, or perturbation to the ball (or table) will be followed by a return to the trajectory. These limit cycles are examples of the attractors previously mentioned, since nearby trajectories (such as those based on different initial conditions) appear to be attracted to them. Figure 9.9 is an example of one complete cycle of trajectories from period-four motion with $\omega = 28.0$ rad/s. None of the four trajectories is the same, but these four repeat exactly, time after time. Since the repetitiveness extends to the table as well as the ball, each period must be an integer multiple of the table's period.

It has been found expedient in nonlinear dynamic analysis not to plot the entire phase plane curve in Fig. 9.8, but rather only a few representative points. For example, the position and velocity of the ball might be plotted every time the table reaches its zero position. For period-one motion, this would be a single point on the phase plane trajectory. For period-two motion, this would correspond to points $P_1$ and $P_2$ such as shown in Fig. 9.8. Since every cycle's trajectory reaches each point at the same relative time, the diagram will consist of only these two points. For period-$N$ motion with this scheme, $N$ points would result. A phase plane plot of these points is called a Poincaré plot and provides an efficient means of illustrating period-$N$ motion. The value of Poincaré plots will become more evident shortly for nonperiodic motion.

The transition from period-$N$ to period $(N + 1)$ motion is important and is demonstrated by returning to the example ($e = 0.9, f = 0.4$, and $A = 0.015$). Figure 9.10 shows a plot of the duration or elapsed time

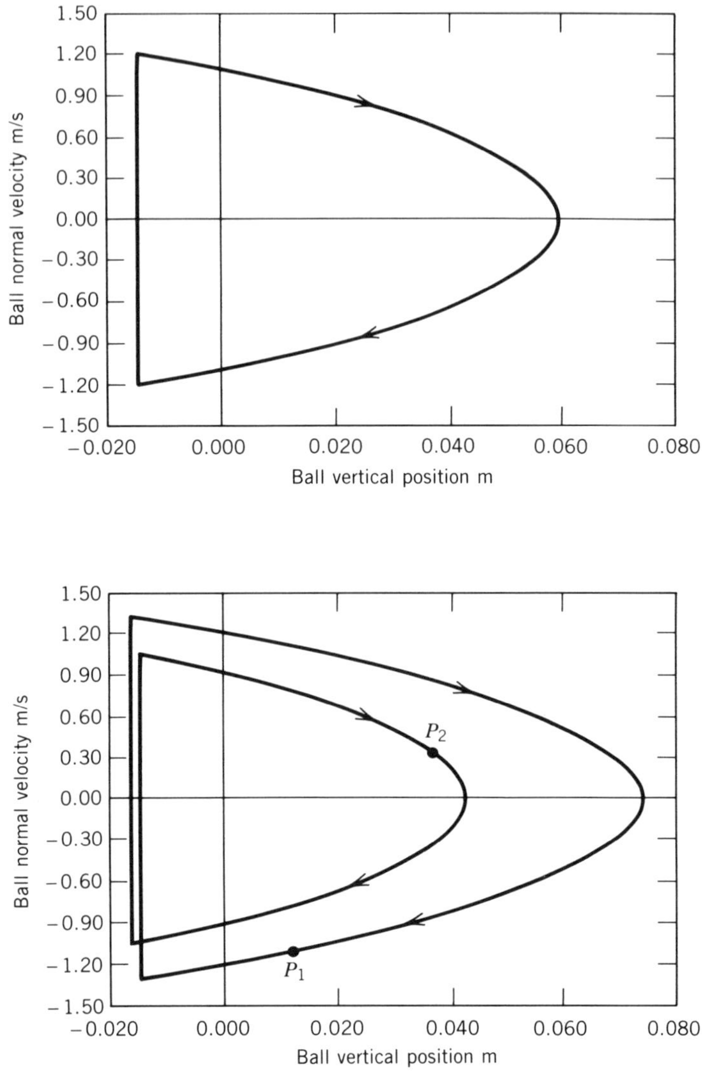


**Figure 9.8.** Upper curve is a closed phase plane trajectory, period-one motion; $\omega = 25.5$ rad/s and $A = 0.015$ m. Lower curve is a closed phase plane trajectory, period-two motion; $\omega = 26.0$ rad/s and $A = 0.015$ m.

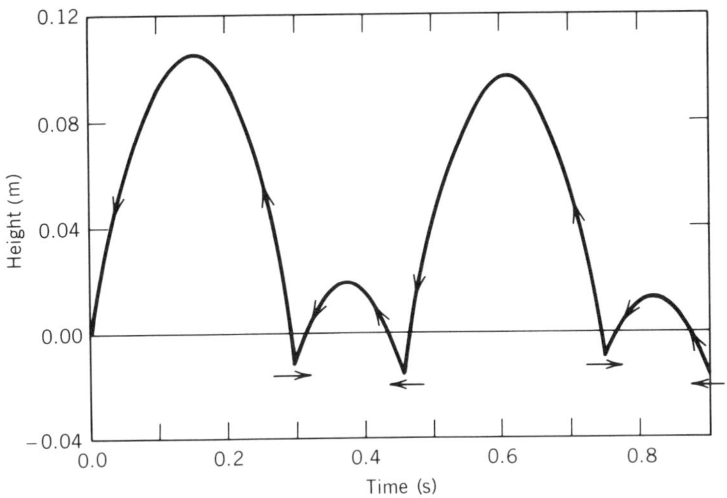


**Figure 9.9.** Period-4 motion; arrows indicate direction of initial relative tangential velocity at impact.

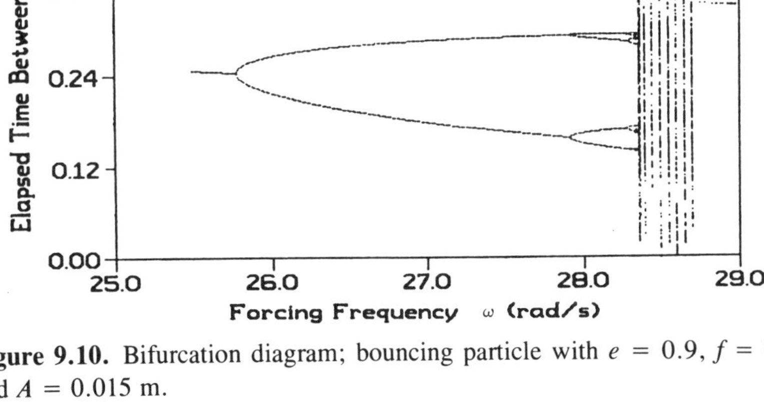


**Figure 9.10.** Bifurcation diagram; bouncing particle with $e = 0.9$, $f = 0.4$, and $A = 0.015$ m.

between impacts for various table frequencies ranging from $\omega = 25.5$ to 29.0 rad/s. The single horizontal line beginning at 25.5 rad/s indicates that only one value of elapsed time between impacts appears in each period of motion. The time interval between bounces remains constant and each postbounce trajectory is identical. As the forcing frequency is increased (in increments of 0.01 rad/s for Fig. 9.10) transition from period-one to period-two motion occurs near $\omega = 25.75$ resulting in two different elapsed times and parabolic trajectories in each period of the motion. The sudden transition from one type of periodic motion to another is called a bifurcation; each curve splits into two branches. Further increase of the forcing frequency results in more bifurcations and higher-order periodic motions. Period-four, period-eight and period-sixteen motions can be seen, after which resolution makes it difficult to identify more complex periodic motions. One of the remarkable features of bifurcation diagrams such as Fig. 9.10 is that if the resolution is increased (that is, the curves are magnified) each new branch keeps splitting or bifurcating into additional pairs.[12] This self-similar geometric structure of the curves is related to the topic of fractals.[10, 19, 20] Increasing the forcing frequency even further results in another region of time intervals in the general vicinity of $\omega = 28.3$ to about $\omega = 28.75$, where the elapsed time curves in Fig. 9.10 eventually lose their identity. More important, the motion of the ball at each table frequency in this region loses its periodicity. Although the physical system is completely deterministic and the table frequency remains constant, the long-term motion never becomes steady or repetitive but is random in appearance. This type of motion is called "chaos." One characteristic of a system capable of chaotic motion is that if a certain parameter is changed by a small amount, the response suddenly can shift from a chaotic form to a periodic one. This can also be seen in Fig. 9.10. Increasing the forcing frequency from 28.7 to 28.75 rad/s produces a period-three motion. Bifurcations again result from further increases in the forcing frequency leading to period-six, period-twelve, and so on, motions and eventually another band of chaos (not shown in Fig. 9.10). In the chaotic regime, small changes in initial conditions lead to widely different trajectories.

As previously seen, the motion of the bouncing ball is nonperiodic for certain ranges of table frequency. If it is not periodic, then it must be irregular. But is it truly random in a probabilistic sense? The answer is no and the Poincaré plot discussed above can be used to demonstrate this. Figure 9.11 shows a Poincaré (phase plane) plot of about 5000 phase points corresponding to the normal velocity after impact and the value of $\omega\tau$ of the table at the time of impact (for $e = 0.9, f = 0.4, A = 0.015$ m, and $\omega = 28.37$ rad/s). If the motion were uniformly random, the phase points would have no discernible pattern and uniformly begin to fill the phase space. Instead, this chaotic response has a more-or-less regular pattern. Recall that the Poincaré plot of period-$N$ motion consists of $N$ points or dots. Viewed as a plot with an unlimited number of dots, it is possible to conclude that the response must have an infinitely long period, that is, aperiodic, or nonperiodic. The results shown in Fig. 9.11 are similar to those of Moon,[4] who does not include friction in a solution of the bouncing ball problem.

Reviewing Fig. 9.10 shows that this vibratory impact response begins with a single impact per cycle at low input (table) frequency. As the table frequency takes on larger values, bifurcations appear and the period of the bouncing motion grows as the number of impacts (bounces) in a cycle increase. The bifurcations have a period doubling effect as $\omega$ grows until a region of chaotic, or nonperiodic, motion takes place. The chaotic region ends and other periodic responses again appear. The period-doubling route to chaos is found in a wide range of nonlinear systems.[4,9] However, there are also other routes that have been found, such as quasi-periodicity and intermittency, for example.[20] Many of these systems do not contain impacts such as beam-buckling problems, fluid motion (including atmospheric dynamics), or heartbeat dynamics. Consequently, vibratory impact problems exhibit behavior seen in many other nonlinear physical systems.

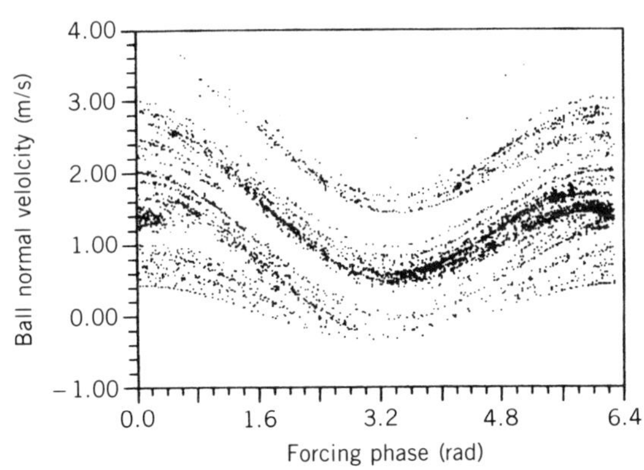


**Figure 9.11.** Poincaré plot; normal velocity sampled at time of impact.

The system parameter varied throughout the previous discussion is the table circular frequency $\omega$. It can be shown[15] that if $\omega$ is held fixed and the table's amplitude $A$ is varied, the same general behavior is observed. That is, as $A$ increases for constant $\omega$, period-doubling bifurcations occur and eventually chaotic regions appear.

Repeating the preceding parameter variations for different values of $e$ produces the same general behavior but changes the values of $\omega$ (or $A$) at which the bifurcations occur. Smaller values of $e$ (higher-energy loss) require higher table frequencies to display the bifurcation process and to reach the chaotic regions.[15] Consequently, varying $e$ (for fixed $\omega$ and $A$) will also show bifurcation behavior. It might be expected that the motions may exhibit some other characteristics when no energy is lost ($e = 1$). Although some analytical simplifications in the problem formulation occur, no different behavior arises in this case.[11]

The effect of varying the one remaining parameter, the friction coefficient, does not affect the bifurcation process and transition to chaos for this problem. This interesting phenomenon occurs because of the algebraic form of the solution and trajectory equations. Recall that in Chapter 2 it was noted that the four system equations for planar particle collisions can be solved sequentially in pairs such that the final normal velocities from an impact are determined independently of the tangential velocities. The tangential velocities are then determined and depend on the normal velocity changes. Note that constant-acceleration particle trajectory equations possess the same property. That is, the normal acceleration is due to gravity and the tangential acceleration is zero (constant velocity); the equations of motion are uncoupled and their solutions are independent. The normal motion determines the height and time of each trajectory independently of the horizontal displacement. This carries over into the mapping equations. The first two mapping equations do not contain the impulse ratio $\mu$ or the tangential velocity components. These two alone form a self-contained or complete nonlinear mapping.[11] Equation 9.39 can be viewed as one that relates the tangential motion to the normal motion, independently determined by the two-equation mapping. Consequently, the presence of friction and tangential motion in the bouncing ball problem does not directly influence the bifurcation process. The reverse is not true, however; periodic or chaotic normal motion will always imply periodic and chaotic tangential motion.[15] This is discussed here for various values of $f$.

In the transient phase of motion (motion prior to reaching a steady state) with relatively high friction as used previously ($f = 0.4$), the tangential velocity of the ball becomes equal to that of the surface after only a few impacts. When steady-state period-one motion develops, the initial and final relative tangential velocities remain zero at each impact, since there are no forces acting on the ball in the tangential direction between impacts. Consequently, the steady-state impulse ratio $\mu$ is zero for all high-friction period-one motion. With low friction, a larger number of impacts is required before the vibrating surface brings the ball's tangential velocity to equal its own, but the end result is the same. For period-two motion, a friction coefficient of 0.4 causes sliding to stop before separation for all impacts. The ball is alternately sped up to the vibrating surface's tangential velocity by the $k$th impact, then it is slowed down to the vibrating surface's tangential velocity by the $(k + 1)$st impact. For $f = 0.4$, period-four, period-eight, and period-sixteen motions all behave in a similar manner. When the friction coefficient is low, sliding can continue through some or all of the impacts. For example, for period-eight motion and for a friction coefficient of $f = 0.05$, sliding stops prior to separation four times out of the eight impacts of the cycle. A set pattern develops as is shown in Table 9.1. For the case of $f = 0.1$, sliding stopped six times out of the eight; another set pattern develops and is shown in Table 9.2. In all cases, however, the periodic nature (period-eight motion) remains unchanged. In all cases, when the normal motion is chaotic, the tangential motion is also.

## 9.4 MASS–SPRING SYSTEMS: IMPACT DAMPERS

The last system to be examined in this chapter is a linear mass–spring system as illustrated in Fig. 9.12 and with an applied harmonic force, $F(\tau)$. Although the spring is linear, the presence of impact velocity discontinuities when the mass strikes the barrier makes the response nonlinear. This system lends itself to a relatively simple analysis, because, between impacts, its response can be found using linear methods. Variations of this problem exist, such as when barriers limit both negative and positive displacements. Systems such as these are often referred to as impact dampers or impact absorbers, since energy is lost, or dissipated, in the absence of viscous and other types of damping. Although they can be noisy, impact dampers have advantages over other types of dampers, such as at very high temperatures when ordinary damping ma-

**TABLE 9.1 Period-Eight Motion, $f = 0.05$**

| $\omega$ (rad/s) | Time between Impacts (s) | Initial Ball Velocity (m/s) | | Final Ball Velocity (m/s) | | Initial Surface Velocity (m/s) | | Impulse Ratio $\mu$ |
|---|---|---|---|---|---|---|---|---|
| | | Normal | Tangent | Normal | Tangent | Normal | Tangent | |
| 28.30 | 0.29185 | −0.7449 | 0.27347 | 1.46688 | 0.16288 | 0.4191 | 0.06690 | −0.05000 |
| 28.30 | 0.17173 | −1.3962 | 0.16288 | 0.82559 | 0.27397 | −0.2268 | 0.35881 | 0.05000 |
| 28.30 | 0.28200 | −0.8591 | 0.27397 | 1.38407 | 0.27721 | 0.3214 | 0.27721 | 0.00144 |
| 28.30 | 0.14240 | −1.3824 | 0.27721 | 0.64444 | 0.28384 | −0.3156 | 0.28384 | 0.00327 |
| 28.30 | 0.29360 | −0.7525 | 0.28384 | 1.47404 | 0.17251 | 0.4193 | 0.06598 | −0.05000 |
| 28.30 | 0.16765 | −1.4062 | 0.17251 | 0.80277 | 0.28296 | −0.2436 | 0.34764 | 0.05000 |
| 28.30 | 0.28555 | −0.8419 | 0.28296 | 1.40297 | 0.25471 | 0.3395 | 0.25471 | −0.01258 |
| 28.30 | 0.14133 | −1.3983 | 0.25471 | 0.64161 | 0.27347 | −0.3246 | 0.27347 | 0.00919 |

**TABLE 9.2 Period-Eight Motion, $f = 0.10$**

| $\omega$ (rad/s) | Time between Impacts (s) | Initial Ball Velocity (m/s) | | Final Ball Velocity (m/s) | | Initial Surface Velocity (m/s) | | Impulse Ratio $\mu$ |
|---|---|---|---|---|---|---|---|---|
| | | Normal | Tangent | Normal | Tangent | Normal | Tangent | |
| 28.30 | 0.29185 | −0.7449 | 0.27347 | 1.46688 | 0.06690 | 0.4191 | 0.06690 | −0.09339 |
| 28.30 | 0.17173 | −1.3962 | 0.06690 | 0.82559 | 0.28908 | −0.2268 | 0.35881 | 0.10000 |
| 28.30 | 0.28200 | −0.8591 | 0.28908 | 1.38407 | 0.27721 | 0.3214 | 0.27721 | −0.00528 |
| 28.30 | 0.14240 | −1.3824 | 0.27721 | 0.64444 | 0.28384 | −0.3156 | 0.28384 | 0.00327 |
| 28.30 | 0.29360 | −0.7525 | 0.28384 | 1.47404 | 0.06598 | 0.4193 | 0.06598 | −0.09784 |
| 28.30 | 0.16765 | −1.4062 | 0.06598 | 0.80277 | 0.28689 | −0.2436 | 0.34764 | 0.10000 |
| 28.30 | 0.28555 | −0.8419 | 0.28689 | 1.40297 | 0.25471 | 0.3395 | 0.25471 | −0.01433 |
| 28.30 | 0.14133 | −1.3983 | 0.25471 | 0.64161 | 0.27347 | −0.3246 | 0.27347 | 0.00919 |

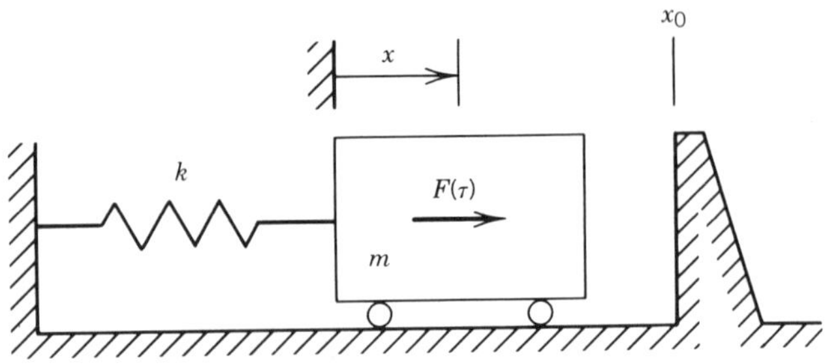


**Figure 9.12.** Linear oscillator model with barrier impact.

terials are not applicable. Another interesting application of the one-sided barrier model[12] is to a ship tethered adjacent to a dock in the presence of wave motion.

Without impact, low-amplitude motion of this system is harmonic at either or both of only two frequencies, the natural frequency $\omega_n$ and the forcing frequency $\omega_f$. For steady-state motion with at least a minimal amount of damping only the latter remains. When impacts are present, it has been known for many years[16, 17] that a large variety of trajectories can exist, some stable and some unstable. More recently, it has been recognized that the response also can be chaotic and the response of this type of system is now being re-examined. A particularly thorough analysis has been presented by Shaw and Holmes[13] who study a more general model than discussed here. The spring in their model has a dual or bilinear stiffness. When the second spring rate is made infinitely large, an impact is simulated. Their results for the impact absorber of Fig. 9.12 are used as a source for much of the following discussion.

The differential equation of motion of the system can be written as

$$\ddot{x} + \omega_n^2 x = \beta \cos \omega_f \tau, \qquad x < 0 \tag{9.40}$$

where the barrier is arbitrarily located at $x_0 = 0$. In Eq. 9.40, $\beta = F/m$ and $\omega_n^2 = k/m$. The general solution of this equation is evaluated following the $k$th impact using the initial conditions $x(\tau_k) = 0$ for the position and

$$\dot{x}(\tau_k^+) = V_n^k = -ev_n^k = -e\dot{x}(\tau_k^-) \tag{9.41}$$

for the velocity. The notation is that $\tau_k^+$ and $\tau_k^-$ are used to indicate "just after'' and ''just before'' the impact at time $\tau_k$. The solution of Eq. 9.40 is[2]

$$x(\tau) = \frac{\beta}{k - m\omega_f^2}(\cos \omega_f \tau - \cos \omega_n \tau) - \frac{e\dot{x}(\tau_k^-)}{\omega_n} \sin \omega_n \tau \quad (9.42)$$

Successive times of impact, $\tau_{k+1}$ are found from Eq. 9.42 when $x$ first returns to $x_0$. As in the previous bouncing ball problem, finding the successive time of impact involves solving a transcendental equation and must be done numerically. The mapping equations in this problem are those that furnish the time and velocity of successive impacts and can be written in a general form as

$$\tau_{k+1} = f(\tau_k, \dot{x}_k) \quad (9.43)$$

and

$$\dot{x}_{k+1} = g(\tau_k, \dot{x}_k) \quad (9.44)$$

The first of these equations is Eq. 9.42 with $x(\tau) = x_0$, suitably solved for the time $\tau_{k+1}$. The second is simply Eq. 9.41, where $\dot{x}(\tau_k^-)$ is found

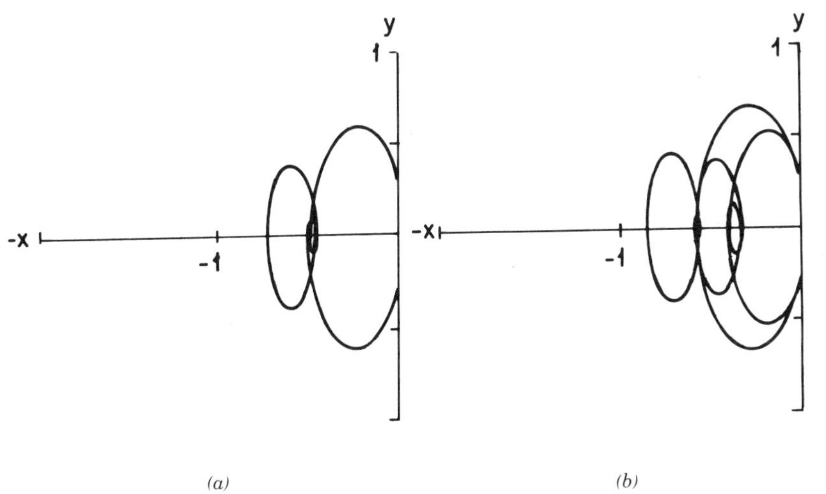


**Figure 9.13.** Example phase plane trajectories with multiple oscillations per impact: (*a*) period-(1,3) motion, (*b*) period-(5,2) motion (from Ref. 13, with permission of Academic Press Limited, copyright 1983).

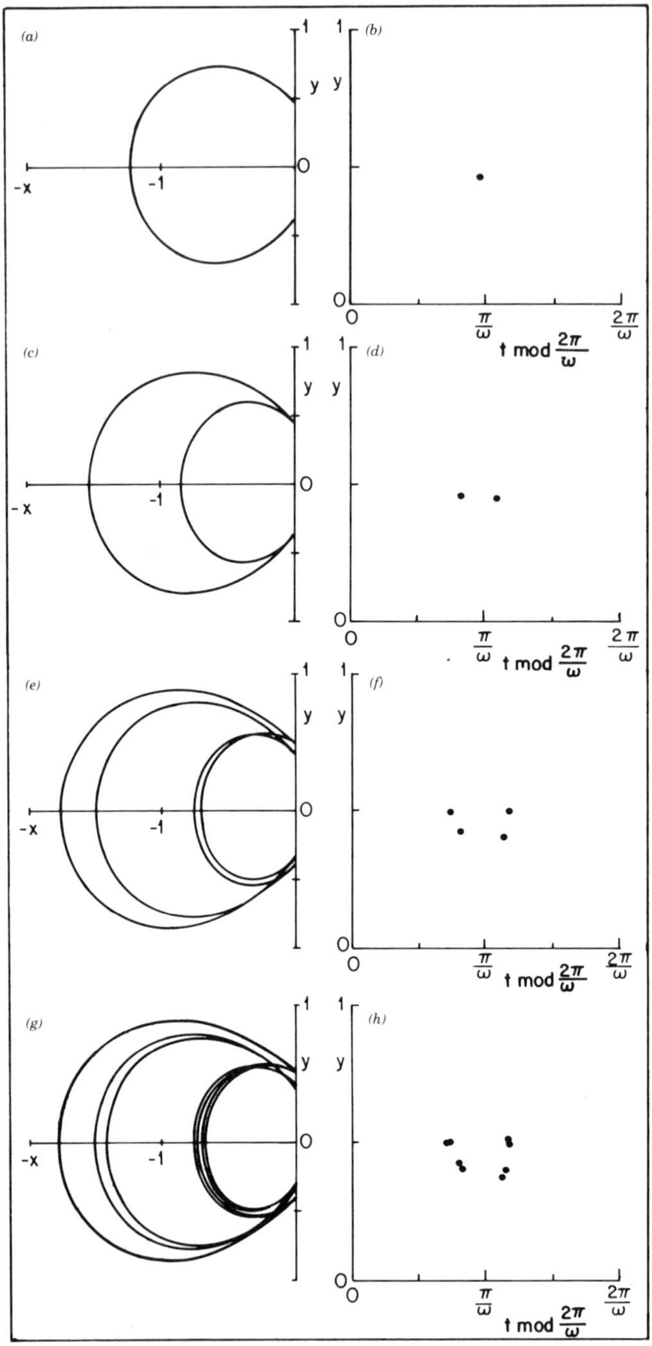


**Figure 9.14.** Sample phase plane trajectories and Poincaré plots: (*a* & *b*) period-(1,1) motion, (*c* & *d*) period-(2,2) motion, (*e* & *f*) period-(4,4) motion, (*g* & *h*) period-(8,8) motion (from Ref. 13, with permission of Academic Press Limited, copyright 1983).

from the derivative of Eq. 9.42. In the bouncing ball problem, periodic motion consisted of closed phase plane trajectories with one or more impacts per cycle, called period-$N$ motion. For the oscillator problem at least one more level of complexity is possible. With the presence of the spring and harmonic forces, full oscillations (phase plane loops) can occur between impacts, so it is possible to have various combinations of $N$ impacts and $M$ oscillations per cycle. Examples are shown in Fig. 9.13. In this context, "period-($N$, $M$) motion" can be used to indicate a cycle with $N$ impacts and $M$ oscillations ($2M$ crossings of the $x$ axis in the phase plane). As a system parameter such as forcing frequency $\omega_f$ is swept through a range of values, different types of motion appear. Figure 9.14 shows the phase plane trajectories and corresponding Poincaré plots for selected values of $\omega_f$ ranging from 2.640 to 2.731 rad/s. These are all period-($N$, $N$) motion. Figure 9.15 shows a cascading set of bifurcations, corresponding to Fig. 9.14. As the system is driven at higher and higher frequencies, bands of chaotic motion occur (shown as open spaces in Fig. 9.15). When chaos ends, the bifurcation process reverses as the periodic motion reverts to simpler, periodic forms. Note that a "transition" is indicated. This is when period-($N$, $M$) motion

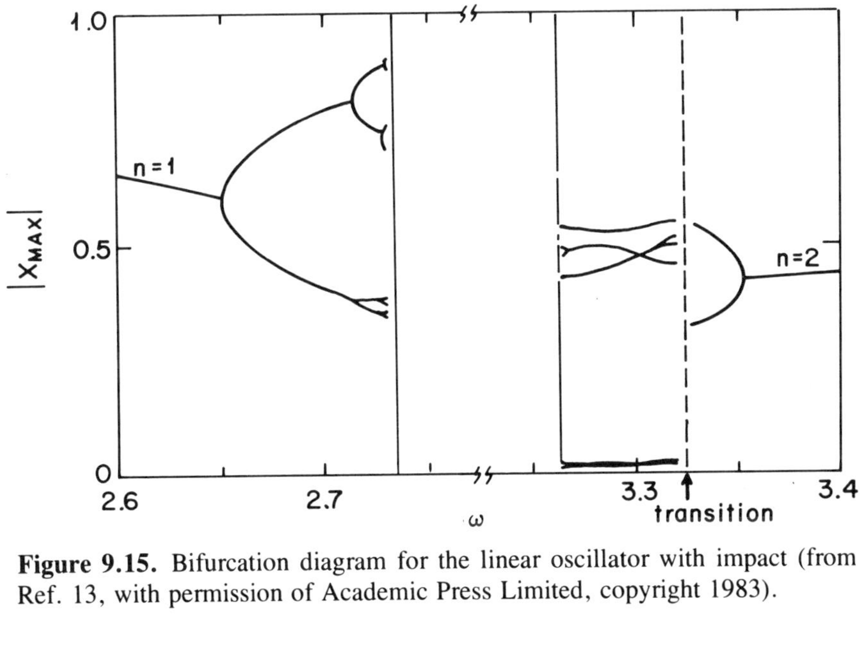


**Figure 9.15.** Bifurcation diagram for the linear oscillator with impact (from Ref. 13, with permission of Academic Press Limited, copyright 1983).

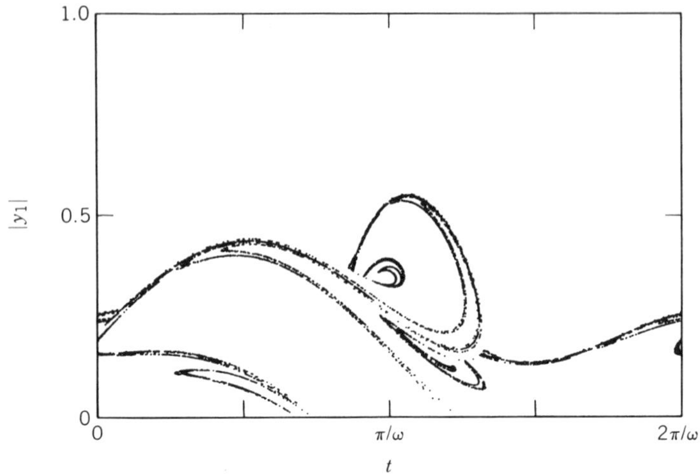


**Figure 9.16.** Poincaré plot for chaotic motion, linear oscillator with impact (from Ref. 13, with permission of Academic Press Limited, copyright 1983).

changes to period-($N - 1$, $M$). This transition is distinguished from a period-doubling bifurcation by Shaw and Holmes,[13] since it is due to the occurrence of a grazing impact, that is, a value of $\dot{x}(\tau_k^-) = 0$ (see Eq. 9.41). This is a type of motion not encountered in the bouncing ball problem although it is physically possible.

In the chaotic bands, the motion is aperiodic and displays distinctive patterns of the Poincaré plots. See Fig. 9.16 for an example. In comparison to limit cycles, these patterns are often called attractors or strange attractors. Isomaki et al.[18] and others have found that these strange attractors can change suddenly; this sudden change is called a chaotic "crisis."

The existence of such a wide variety of motion for vibratory systems accompanied by impacts (and other nonlinear systems) presents a serious challenge to a designer where these impacts are used or at least tolerated. An even greater challenge is to exploit these motions.

## 9.5 PROBLEMS

1. Derive Eq. 9.19 using energy concepts and the condition of steady-state motion.

**2.** Show that Eq. 9.17 can be written as

$$V_{n(i+1)}^2 = e^4 V_{ni}^2 + (1 - e^4) V_{ns}^2$$

Using this equation, perturb $V_{n(i+1)}^2$ and $V_{ni}^2$ by the amounts $\delta V_{n(i+1)}^2$ and $\delta V_{ni}^2$, respectively. Show that for $e < 1$ the perturbation dies out, that is, show that motion is asymptotically stable.

## REFERENCES


1. Timoshenko, S., Young, D. H., and Weaver, Jr., W., *Vibration Problems in Engineering*, 4th ed., John Wiley, New York, 1974.
2. Steidel, Jr., R. F., *An Introduction to Mechanical Vibrations*, John Wiley, New York, 1989.
3. Thompson, J. M. T. and Stewart, H. B., *Nonlinear Dynamics and Chaos*, John Wiley, New York, 1986.
4. Moon, F. C., *Chaotic Vibrations*, John Wiley, New York, 1987.
5. Colver, G. M., "Dynamic and stationary charging of heavy metallic and dielectric particles against a conducting wall in the presence of a dc applied electric field," *J. Appl. Phys.*, **47**(11), 4839–4849, 1976.
6. Olansen, J. B. and Dunn, P. F., "Dispensing particles under atmospheric and vacuum conditions using an electrostatic device," *J. Appl. Phys.*, **66**(12), 6098–6109, 1989.
7. Stoker, J. J., *Nonlinear Vibrations*, J. Wiley, New York, 1966.
8. Akashi, H., "The motion of an electric bell," Am. Math. Monthly, **65,** 255–258, 1958.
9. Thompson, J. M. T. and Stewart, H. B., *Nonlinear Dynamics and Chaos*, John Wiley, New York, 1986.
10. Jensen, R. V., "Classical chaos," *American Scientist*, **75,** March–April, p. 168–181, 1987.
11. Holmes, P. J., "The Dynamics of Repeated Impacts with a Sinusoidally Vibrating Table," *J. Sound Vibr.*, **84**(2), pp. 173–189, 1982.
12. Thompson, J. M. T. and Ghaffari, R., "Chaos after period-doubling bifurcations in the response of an impact oscillator," *Phys. Lett.*, **91A**(1), 4–8, 1982.
13. Shaw, S. W. and Holmes, P. J., "A periodically forced piecewise linear oscillator," *J. Sound Vibr.*, **90**(1), 129–155, 1983.

14. Tufillaro, N. B. and Albano, A. M., "Chaotic dynamics of a bouncing ball," *Am. J. Phys.*, **54**(10), 939–944, 1986.
15. Kozol, J. E., *Two-Dimensional Vibratory Impact with Chaos*, MS Thesis, University of Notre Dame, 1989.
16. Lieber, P. and Jensen, D. P., "An acceleration damper: Development, design and some applications," *Trans. ASME*, **67,** 523–530, 1945.
17. Masri, S. F. and Caughey, T. K., "On the stability of the impact damper," *Trans. ASME, J. Appl. Mech.*, 586–592, 1966.
18. Isomaki, H. M., Von Boehm, J. and Räty, R., "Devil's attractors and chaos of a driven impact oscillator," *Phys. Lett.*, **107A**(8), 343–346, 1985.
19. Mandelbrot, B. B., *The Fractal Geometry of Nature*, W. H. Freeman, New York, 1977.
20. Bergé, P., Pomeau, Y. and Vidal, C., *Order Within Chaos*, John Wiley, New York, 1984.

# CHAPTER 10

# APPLICATION TO VEHICLE COLLISIONS

Probably the most commonplace of collisions that directly affect people is that of vehicles, from bicycles to aircraft. Many of these collisions meet the assumptions made in the development of the system equations in this book and are amenable to this approach. The need for analysis of vehicle collisions is varied. Safety studies have noted a correlation between vehicle occupant injury severity and velocity change, $\Delta V$, where $\Delta V$ is the magnitude of the velocity change vector. Settlement of insurance claims and litigation and criminal proceedings frequently need information on accident causation or preimpact vehicle speeds or both. Vehicle component failures with accident causation potential, occasionally require collision analyses for damage assessment. Most of the applications of collision analysis have been to automobiles and trucks.[1–7] Potential exists for analysis of some aircraft crashes as a result of experiments carried out for aircraft-to-ground collsions,[8] but more research remains to be done on this topic.

Vehicle collision studies in general can be placed broadly into two categories: (1) structural crush analysis and design and (2) velocity change ($\Delta V$) calculations. The former use analytical tools of a structural engineer, such as theories of elasticity and plasticity along with finite element analysis to perform vehicle structural design. This is a complex problem and, although related to $\Delta V$ calculation, will not be covered here. Methods for estimating collision velocity changes, frequently related to a specific collision, are discussed here. This topic is part of what is referred to as accident reconstruction. Only planar collisions are considered. Most of the chapter deals with ground vehicles, primarily automobiles. Some discussion of using impulse and momentum concepts for collisions of light aircraft with the ground is included. Application techniques of three-dimensional theory to vehicle collisions have yet to be developed.

Two basic approaches have developed for collision analysis of ground vehicles. One uses the planar solution equations directly, including the coefficients. Using various strategies, this approach attempts to match calculated impact velocities to existing empirical information[5,13] for a given accident. Another approach uses a limited form of the solution equations (primarily the point-mass solution) combined with an estimate of energy loss obtained by measuring structural deformation.[1–3,12] In this approach, the measurements of the permanent deformation or crush on a given vehicle are compared to a data base that provides a "crush energy loss." This energy loss together with the point-mass solution is used to estimate $\Delta V$. Since energy loss is directly related to the coefficients $e$ and $\mu$, this approach is not fundamentally different from direct application of the solution equations. The accuracy of the method of transforming the crush measurements to energy loss through a data base has been questioned,[9,10] but is under continuing development.[27] The discussion here centers on the direct application of the planar collision solution equations to vehicle collisions. Some comparisons of the approaches are made. Combinations of these methods currently are being investigated.

The application of impulse and momentum equations to collision analysis is not new.[16] Much of the earlier work was limited to frontal crashes at moderate to high speeds, where the coefficient of restitution is near zero and the coefficient of "friction" does not enter into the problem. Later when more complicated collision configurations were analyzed, attempts were made to determine values of the coefficients. Early estimates[16,17] of $\mu$ centered around 0.4 but went as high as "infinity."[18] This is not as unreasonable as it sounds, since it is likely that a large value of a friction coefficient was sought to ensure that sliding ended before the collision was over, which is a common occurrence for a large class of automobile collisions. Of course, there is no reason to expect that the forces tangent to the crush surfaces of colliding vehicles are due exclusively or even predominantly to dry friction. Interference and interlocking of parts commonly occurs. This is a good example of why the concept of impulse ratio and its limiting value, rather than simply a friction coefficient is necessary. This implies, however, that experimental data must be used to find $\mu$ for some collisions.

A big advance in the use of impulse and momentum analysis of vehicle collisions was the result of a particular group of experimental collisions. Controlled collisions have been conducted since the 1950s and occur now worldwide almost on a daily basis. However, a large majority of these are frontal barrier collisions, and moving-barrier side collisions conducted by manufacturers, very special collision configurations. The breakthrough came with a group of collisions known by the acronym RICSAC.[11] The reason these crashes are special is because they were the first set of planned, comprehensive crashes instrumented to measure not only linear but also angular velocities and postimpact motion (motion to the final resting positions). Data from these collisions have been analyzed to obtain estimates of the coefficients $e$ and $\mu$.[13]

The RICSAC experiments and their major features are described in the second section of this chapter. First, however, the planar rigid body system equations are reformulated and presented. Their form here includes the use of a scene-based $x$-$y$ coordinate system, which is often convenient in accident analysis. The third section summarizes the results of fitting the RICSAC data using the method of least squares to the planar model solution equations. In effect, this provides experimental values for the coefficients. The remainder of this chapter includes several topics related to automobile collision analysis and reconstruction, and ends with discussion of aircraft-to-ground collisions. Collision mechanics of articulated vehicles has been discussed in Chapter 8; no experimental data are known to exist for these collisions.

## 10.1 SCENE-BASED PLANAR SYSTEM EQUATIONS

Figure 10.1 shows free body diagrams of two colliding vehicles. In contrast to Chapter 6, the system equations are summarized here in a form where all of the variables are expressed in terms of the scene coordinates $x$ and $y$ and velocities rather than the impact surface coordinates, $n$ and $t$. The normal and tangential coordinates still enter the problem implicitly since the effects characterized by $e$ and $\mu$ lie perpendicular to and along these axes. The angle $\Gamma$ relates the two coordinate systems.

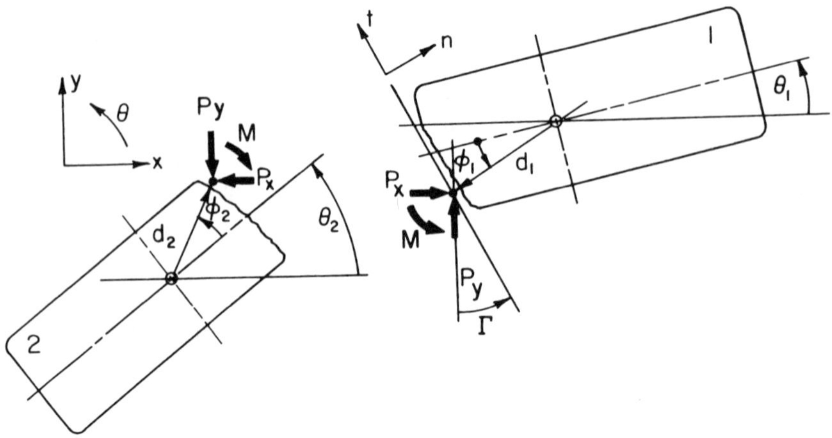


**Figure 10.1.** Free body diagrams of colliding vehicles.

The system equations and their source are summarized as:

- Conservation of linear momentum, $x$ axis:

$$m_1 V_{1x} + m_2 V_{2x} = m_1 v_{1x} + m_2 v_{2x} \tag{10.1}$$

- Conservation of linear momentum, $y$ axis:

$$m_1 V_{1y} + m_2 V_{2y} = m_1 v_{1y} + m_2 v_{2y} \tag{10.2}$$

- From restitution, normal to the impact surface:

$$\begin{aligned}(V_{1y} - d_d\Omega_1 - V_{2y} - d_b\Omega_2)\sin\Gamma \\ + (V_{1x} + d_c\Omega_1 - V_{2x} + d_a\Omega_2)\cos\Gamma \\ = -e\big[(v_{1y} - d_d\omega_1 - v_{2y} - d_b\omega_2)\sin\Gamma \\ + (v_{1x} + d_c\omega_1 - v_{2x} + d_a\omega_2)\cos\Gamma\big]\end{aligned} \tag{10.3}$$

where

$$\begin{aligned} d_a &= d_2 \sin(\theta_2 + \phi_2), \qquad d_c = d_1 \sin(\theta_1 + \phi_1) \\ d_b &= d_2 \cos(\theta_2 + \phi_2), \qquad d_d = d_1 \cos(\theta_1 + \phi_1) \end{aligned} \tag{10.4}$$

- Impulse ratio equation:

$$m_1(V_{1y} - v_{1y})(\cos \Gamma - \mu \sin \Gamma) + m_2(V_{2x} - v_{2x})(\sin \Gamma + \mu \cos \Gamma) = 0 \quad (10.5)$$

- Conservation of angular momentum of the system:

$$I_1(\Omega_1 - \omega_1) + I_2(\Omega_2 - \omega_2) - m_2(d_a + d_c) \cdot (V_{2x} - v_{2x}) + m_1(d_b + d_d)(V_{1y} - v_{1y}) = 0 \quad (10.6)$$

- Moment restitution:

$$(\Omega_2 - \Omega_1)(1 + e_m) = -e_m[(\Omega_1 - \omega_1) - m_1 d_c(V_{1x} - v_{1x})/I_1 + m_1 d_d(V_{1y} - v_{1y})/I_1 - (\Omega_2 - \omega_2) - m_2 d_a(V_{2x} - v_{2x})/I_2 + m_2 d_b(V_{2y} - v_{2y})/I_2] \quad (10.7)$$

As earlier, when $e_m = -1$, the moment impulse is zero; otherwise a rotationally elastic impact occurs for $0 \leq e_m \leq 1$.

These equations are most easily solved using a linear equation solution routine and a digital computer. All solutions presented in this chapter were so obtained. It should be pointed out that for many vehicular accident analysis problems, the final velocity components are often the "knowns" and the initial velocity components are the "unknowns." This is referred to as the inverse solution. The form of Eq. 10.1–10.7 permits an inverse solution to be found, but the inverse solution is singular for $e = 0$ or $e_m = 0$ and does not exist.

## 10.2 EXPERIMENTAL COLLISION DATA (RICSAC Data)

Twelve, two-car staged collisions were conducted[11] for the U.S. National Highway Traffic Safety Administration. One of the collisions suffered a data loss; only 11 are discussed in this chapter. Other collisions were conducted subsequently,[12] but their data are not included here. Claims have been made that some errors exist in the original data,[9] but the data used here are directly from Ref. 11 with corrections only for accelerometer location as described in Ref. 13.

Figure 10.2 shows the four collision configurations that were used in the RICSAC collisions. A mixture of vehicles was used in combination with different preimpact speeds. Table 10.1 lists all of the pertinent data corresponding to the 11 collisions, including the measured final velocities. Vehicles in all collisions had zero initial angular velocities. It should be pointed out that for purposes of analysis, the distances $d_i$ from the mass center to the impulse "point" and the corresponding angles $\phi_i$ for each vehicle were chosen by locating the point visually. The technique used was to choose the impulse point from a plan view of the vehicle's crush profile. Some of the consequences of this subjective choice are discussed in Ref. 13.

If the coefficients $e$, $e_m$, and $\mu$ were known for a given collision, the initial velocities, physical constants, these coefficients, and the solution of the equations from the last section would provide final velocities. These final velocities would be exactly those measured and listed in Table 10.1 except for

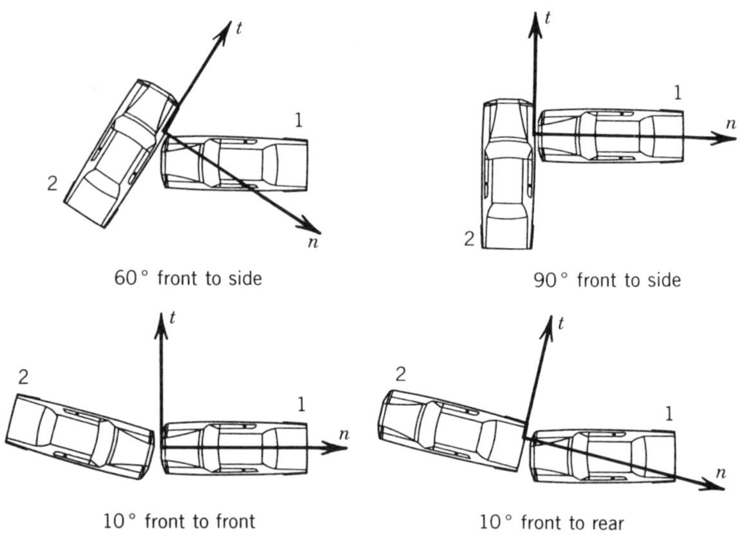


**Figure 10.2.** Initial vehicle configurations for experimental collisions.[11]

**TABLE 10.1 Experimental Collision/Vehicle Data (U.S. units)**

| Collision Number/Vehicle | Collision Category | Vehicle Mass ($lb\text{-}s^2/ft$) | Moment of Inertia ($ft\text{-}lb\text{-}s^2$) | Distance d (ft) | Angle $\phi$ (deg) | Angle of Veh. $\theta$ (deg) | Crush Angle $\Gamma$ (deg) | Initial Velocity (ft/s) $v_x/v_y$ | Final Angular Velocity (rad/s) | Final Velocity (ft/s) $V_x/V_y$ |
|---|---|---|---|---|---|---|---|---|---|---|
| 1/1 | 60° | 143.5 | 3728 | 7.59 | −19.8 | 0 | −30 | −29.04/0.0 | −1.57 | −12.33/7.91 |
| 1/2 | Front | 95.8 | 1961 | 3.44 | −38.7 | 60 | −30 | 14.52/25.15 | 0.0 | −6.80/16.97 |
| 6/1 | To | 133.6 | 3469 | 8.41 | −17.9 | 0 | −30 | −31.53/0.0 | −0.52 | −18.68/4.12 |
| 6/2 | Side | 81.5 | 1669 | 2.00 | −90.0 | 60 | −30 | 15.29/27.31 | −3.14 | −4.20/18.02 |
| 7/1 | | 115.0 | 2985 | 8.41 | −17.9 | 0 | −30 | −42.68/0.0 | −0.52 | −25.41/4.85 |
| 7/2 | | 81.1 | 1082 | 2.00 | −90.0 | 60 | −30 | 21.34/36.96 | −3.35 | −7.63/28.34 |
| 8/1 | 90° | 139.1 | 3614 | 7.90 | 0.0 | 0 | 0 | −30.51/0.0 | −1.99 | −10.24/10.72 |
| 8/2 | Front | 146.3 | 3800 | 2.77 | −68.8 | 90 | 0 | 0.0/30.51 | −0.31 | −12.02/19.72 |
| 9/1 | To | 70.1 | 976 | 4.80 | 6.0 | 0 | 0 | −31.09/0.0 | −3.14 | −2.81/14.83 |
| 9/2 | Side | 152.2 | 3953 | 5.20 | −29.7 | 90 | 0 | 0.0/31.9 | 0.79 | −9.90/24.20 |
| 10/1 | | 71.6 | 998 | 3.47 | 0.0 | 0 | 0 | −48.84/0.0 | −5.24 | −5.08/28.18 |
| 10/2 | | 146.6 | 3008 | 5.29 | −29.2 | 90 | 0 | 0.0/48.84 | 1.26 | −14.57/36.55 |
| 11/1 | 10° | 94.4 | 1935 | 6.14 | 9.4 | 0 | 0 | −29.92/0.0 | 0.52 | 5.81/2.02 |
| 11/2 | Front | 150.6 | 3913 | 7.66 | 11.3 | −10 | 0 | 29.47/−5.2 | 0.0 | 6.42/−4.11 |
| 12/1 | To | 97.2 | 1992 | 5.90 | 9.6 | 0 | 0 | −46.20/0.0 | 1.57 | 14.05/−1.61 |
| 12/2 | Side | 140.2 | 3640 | 7.28 | 10.3 | −10 | 0 | 45.50/−8.02 | 1.05 | 6.32/−9.65 |
| 3/1 | 10° | 153.7 | 3992 | 8.83 | −17.0 | 0 | −10 | −31.09/0.0 | −0.26 | −17.15/0.24 |
| 3/2 | Front | 97.0 | 1986 | 7.63 | 171.4 | 170 | −10 | 0.0/0.0 | 0.0 | −22.87/3.74 |
| 4/1 | To | 154.7 | 4018 | 8.02 | −18.2 | 0 | −10 | −56.76/0.0 | −0.65 | −29.33/−1.43 |
| 4/2 | Rear | 99.1 | 2032 | 6.94 | 171.7 | 170 | −10 | 0.0/0.0 | −0.52 | −32.54/1.38 |
| 5/1 | | 142.9 | 3711 | 8.08 | −20.7 | 0 | −10 | −58.23/0.0 | −0.21 | −34.31/0.57 |
| 5/2 | | 78.6 | 1095 | 5.75 | 168.0 | 170 | −10 | 0.0/0.0 | −1.22 | −37.15/2.77 |

1. experimental error,
2. improper choice of $d$ and $\phi$,
3. errors in physical property values of the vehicles,
4. errors in the choice of an equivalent impact surface, that is, choice of $\Gamma$,
5. violation of assumptions.

Unfortunately, none of the RICSAC experiments were repeated under identical conditions so no direct measure of experimental error is available. Consequently, the sources of errors are confounded. Because of the error, each data set in Table 10.1 could not be expected to satisfy the system or solution equations exactly. If the physical parameters are assumed to be accurate and the measured final velocities in error, then a nearby set of values of the final velocities will satisfy the equations. What is done is to find the coefficients $e$, $e_m$ and $\mu$ that provide these "nearby" velocity values in a least square sense.

## 10.3 LEAST SQUARE ANALYSIS

A summary is presented here of the results of the process of fitting the system equations to the data of Table 10.1 using the least square approach. Details are contained in Refs. 13 and 14. A sum of squares of differences was formed of the experimentally measured final velocity components $V_{1x}$, $V_{1y}$, $V_{2x}$, $V_{2y}$, $\Omega_1$, and $\Omega_2$ and the corresponding values of the same variables that satisfy the system equations. That is,

$$Q = w_1(V_{1x} - \hat{V}_{1x})^2 + w_2(V_{1y} - \hat{V}_{1y})^2 + w_3(V_{2x} - \hat{V}_{2x})^2 + w_4(V_{2y} - \hat{V}_{2y})^2 + w_5(\Omega_1 - \hat{\Omega}_1)^2 + w_6(\Omega_2 - \hat{\Omega}_2)^2 \quad (10.8)$$

where a component value with the caret represents the value that satisfies the solution equations. The $w$'s are weighting constants (chosen to give all terms in Eq. 10.8 generic units of energy). The quantity $Q$ was minimized (numerically) with the constraint that the careted variables satisfy the system equations. Minimization was with respect to the coefficients, which were furnished along with the careted velocity components.

### 10.3.1 Coefficients

In order to apply the equations of collision theory to a vehicle collision analysis, appropriate values of the coefficients $e$, $e_m$, and $\mu$ must be known. The values obtained from analysis of the RICSAC collisions are now discussed. Crush surfaces are large. It is certainly possible that moment effects may occur. On the other hand, the concept of a moment coefficient is new; previous analyses of collisions assumed a zero moment impulse. For this reason two least squares solutions were found. One set is for the moment coefficient constrained to $-1$ (zero moment impulse), the other where $e_m$ could take on any value within $0 \leq e_m \leq 1$. The results of both analyses are given in Table 10.2.

One measure of the goodness of fit between the data and equations is the value of $Q$. It is apparent from Table 10.2 that $Q$ is smaller in some cases where a moment impulse is allowed and smaller in other cases where $e_m = -1$. Collisions 9 and 10 show $Q$ to be nine to seven times smaller for $e_m = -1$ than for arbitrary $e_m$. Since a moment impulse can arise through inaccurate selection of the impulse point (i.e., through inaccurate selection of $d$ and $\phi$), as well as due to the physical presence of a moment impulse, the data do not provide an unequivocal conclusion on the existence or lack of existence of a significant moment impulse. However, note that with the exception of collisions 9 and 10, little difference occurs in energy loss between the two analyses. On the basis of these observations and other comparisons not discussed here, it seems that a significant moment impulse is not indicated for these collisions. Consequently, $e_m = -1$ is likely to be the appropriate value for these types of vehicle collisions. This is assumed to be the case for the remaining discussion.

The restitution coefficients seem to vary only a small amount within each collision category. An exception is collision 8 where $e = 0.079$, whereas $e$ for collisions 9 and 10 is 0.4 or higher. It is common to make the approximation that $e = 0$ for collision analysis.[5,6] Based on the RICSAC data, this assumption appears inappropriate for all collisions.

The impulse ratios (or equivalent coefficients of friction) display some interesting properties. They vary only a relatively small amount within each collision category but noticeably between categories. *Every value of $\mu$ in Table 10.2a is equal to or very close to the value of $\mu_0$ for each collision.* This is so consistent—it seems to imply *that for these types of collisions*—it may be concluded that sliding terminates at or prior to

**TABLE 10.2 Results, Least Square Analysis of Experimental Collisions**

| RICSAC COLLISION | 1 | 6 | 7 | 8 | 9 | 10 | 11 | 12 | 13 | 14 | 15 |
|---|---|---|---|---|---|---|---|---|---|---|---|
| | | | | | *a. Zero Moment Impulse, M* | | | | | | |
| Mimimum sum of squares | 55.4 | 198.2 | 221.5 | 61.3 | 29.8 | 100.6 | 21.8 | 75.0 | 29.2 | 96.2 | 115.6 |
| Energy loss (%) | 5.20 | 48.2 | 48.8 | 36.0 | 28.9 | 31.0 | 90.9 | 91.9 | 34.2 | 36.3 | 32.0 |
| Restitution coefficient | 0.000 | 0.000 | 0.000 | 0.079 | 0.400 | 0.419 | 0.000 | 0.100 | 0.217 | .045 | 0.053 |
| Moment coefficient | −1 | −1 | −1 | −1 | −1 | −1 | −1 | −1 | −1 | −1 | −1 |
| Impulse ratio | .966 | .824 | .772 | .413 | .486 | .590 | .038 | .031 | −.065 | −.050 | −.090 |
| | | | | | *b. Nonzero Moment Impulse, M* | | | | | | |
| Mimimum sum of squares | 95.5 | 101.7 | 203.3 | 77.6 | 275.0 | 750.7 | 10.2 | 31.7 | 19.2 | 85.7 | 106.5 |
| Energy loss (%) | 58.2 | 48.2 | 48.8 | 39.5 | 42.4 | 43.5 | 92.2 | 93.3 | 34.1 | 36.3 | 32.3 |
| Restitution coefficient | .000 | .000 | .000 | .034 | .091 | .133 | .000 | .102 | .223 | .045 | .056 |
| Moment coefficient | .000 | .997 | .815 | .000 | .000 | .000 | .000 | .012 | .747 | .517 | .001 |
| Impulse ratio | .922 | .872 | .778 | .500 | .809 | .837 | .025 | .020 | −.052 | −.042 | −.071 |

separation. This has been a commonly made assumption in collision analysis.[5,6] Note, however, that the RICSAC experiments did not contain a collision configuration of the type typically referred to as a sideswipe collision. For this type, by definition, sliding does not end. Consequently, $\mu = \mu_0$ does not apply to all vehicle collision categories.

### 10.3.2 Energy Loss

Although energy loss is not used explicitly in the solution equations, it is a fundamental parameter in collision analysis. As mentioned earlier, empirically based values of energy loss are used directly in some $\Delta V$ methods.[2] Recall also that observation of energy loss leads to the concept of critical impulse ratio. For these reasons, the RICSAC experimental energy loss values are analyzed and compared.

As do the coefficients $e$ and $\mu$, the energy losses in Table 10.2 consistently cluster according to collision category. The average energy loss for each of the four collision categories from Table 10.2a are 49.7%, 32.0%, 91.4%, and 34.2%, respectively. Figure 10.3 illustrates the ratio of energy loss found by the least square method to the experimental values. The least square energy loss is broken into the portions normal and tangential to the crush surface.* In addition, estimates of the energy lost due to tire/ground friction during the contact duration (usually neglected) are added. Frictional energy losses are based on point-mass theory, Chapter 2, for a typical contact duration of 0.15 s and a tire coefficient of friction of 0.87, measured during the RICSAC collisions.

The largest total energy error magnitude is about −30%, collision 4; the next largest are collisions 8 and 5 with errors of −23%. Recall that these errors are due to combinations of measurement error and analysis error. Examples of the latter include errors in the location of the point of impact and impact surface and violation of assumptions. Another example of analysis error is the neglecting of three-dimensional effects. Although smaller errors are desirable, these values are significantly

*Partitioning of the energy loss into a portion associated with normal deformation (restitution $e$) and a portion associated with tangential effect $\mu$ is a convenient and informative concept.[4] It was done in Chapter 3 for point-mass collisions and in Chapter 6 for rigid body barrier collisions. For any collision, the system equations are solved for the appropriate values of $e$ and $\mu$. An additional solution for $\mu = 0$ provides the energy loss associated with $e$; the difference in the energy losses is that associated with $\mu$. See Ref. 6.

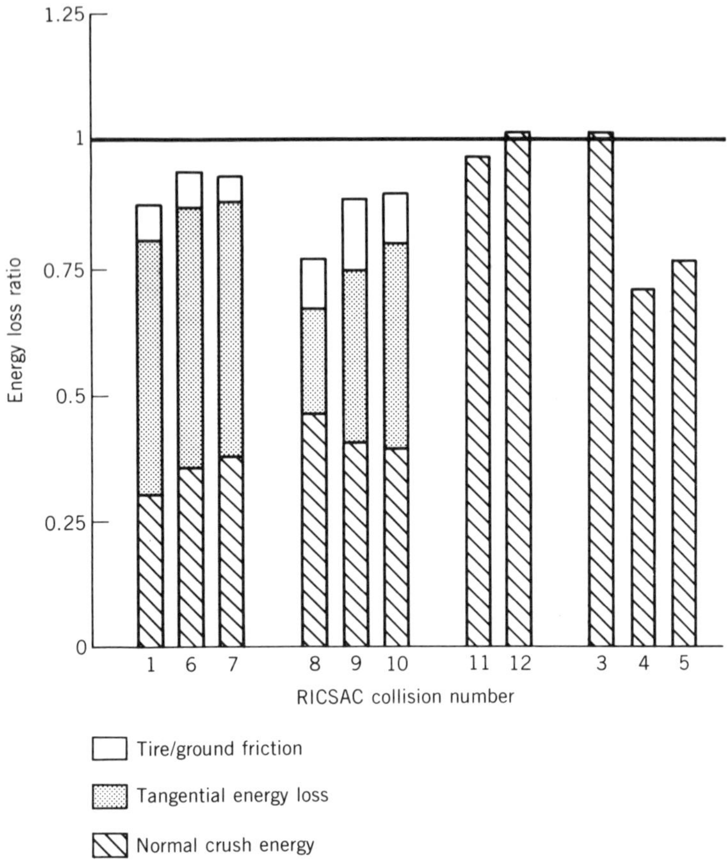


**Figure 10.3.** Energy loss in experimental collisions; from least square data fit.

smaller than other $\Delta V$ models mentioned earlier, which are based on estimation of energy loss on the basis of crush depth measurements.[10]

### 10.3.3 Velocity Change

A remaining comparison to be made concerns how closely the velocity changes from the least square analysis compare to the experimental values. Values of the individual initial and final velocity components for each vehicle could be compared, but the $\Delta V$'s give good indications of overall accuracy. Individual comparisons can be found in Ref. 13. Table 10.3 lists three values of $\Delta V$ for each vehicle corresponding to the 11

**TABLE 10.3 Comparisons of Velocity Changes, RICSAC Collisions**

| Collision/ Vehicle Numbers | Experimental Value (ft/s) | Least Squares (ft/s) | Difference, Experimental minus Least Square | Empirical Value (ft/s) |
|---|---|---|---|---|
| 1/1 | 18.5 | 16.1 | 2.4 | 17.6 |
| 1/2 | 22.8 | 24.2 | −1.4 | 26.4 |
| 6/1 | 13.5 | 14.7 | −1.2 | 17.1 |
| 6/2 | 21.6 | 24.1 | −2.5 | 28.1 |
| 7/1 | 17.9 | 20.4 | −2.5 | 24.1 |
| 7/2 | 30.2 | 28.9 | 1.3 | 34.2 |
| | | | Avg. = −0.65 | |
| | | | St. Dev. = 2.04 | |
| 8/1 | 22.9 | 19.3 | 3.6 | 18.4 |
| 8/2 | 16.2 | 18.4 | −2.2 | 17.5 |
| 9/1 | 31.9 | 32.0 | −0.1 | 31.3 |
| 9/2 | 12.1 | 14.7 | −2.6 | 14.4 |
| 10/1 | 52.0 | 51.6 | 0.4 | 47.8 |
| 10/2 | 19.1 | 25.2 | −6.1 | 23.3 |
| | | | Avg. = −1.17 | |
| | | | St. Dev. = 3.28 | |
| 11/1 | 35.8 | 36.6 | −0.8 | 36.0 |
| 11/2 | 23.1 | 22.9 | 0.2 | 22.6 |
| 12/1 | 60.3 | 59.3 | 1.0 | 52.1 |
| 12/2 | 39.2 | 41.4 | −2.2 | 36.1 |
| | | | Avg. = −0.45 | |
| | | | St. Dev. = 1.38 | |
| 3/1 | 13.9 | 14.1 | −0.2 | 17.1 |
| 3/2 | 23.2 | 22.4 | 0.8 | 19.2 |
| 4/1 | 27.5 | 22.4 | 5.1 | 22.7 |
| 4/2 | 32.6 | 34.9 | −2.3 | 35.5 |
| 5/1 | 23.9 | 20.5 | 3.4 | 22.0 |
| 5/2 | 37.3 | 37.3 | 0.0 | 40.0 |
| | | | Avg. = 1.13 | |
| | | | St. Dev. = 2.68 | |

RICSAC collisions. These are the experimental values, the values from the least square analysis, and those from an empirical equation discussed in the next section. Although scanning the individual differences in the table is informative, some statistics are also included. The average difference between the least square fit and experimental values ranges from −1.17 to 1.13 ft/s for the four collision categories. A few individual $\Delta V$'s varied considerably, but the averages are reasonably small in all collision categories. Note that $\Delta V$ errors are generally smaller than errors in energy loss. This is because velocity is proportional to the square root of energy.

## 10.4 EMPIRICAL EQUATION FOR $\Delta V$

Following the analysis of the RICSAC collisions, it was noted that an empirical relationship exists between each vehicle's $\Delta V$ and the initial momentum and energy loss of the collision. The following equation furnishes an estimate of $\Delta V$ for each vehicle:

$$\Delta V_i = (\Delta T)^{1/2}[(m_1 v_1)^2 + (m_2 v_2)^2]^{1/2}/m_i; \qquad i = 1, 2 \quad (10.9)$$

The subscript $i$ represents either vehicle and $\Delta T$ is the fractional energy loss in the collision, $T_L/T_{\text{init}}$. The last three columns of Table 10.3 contain the values of each vehicle's velocity change from experimental measurements, the least square fitting of the impact equations and the value predicted by Eq. 10.9, respectively. Figure 10.4 illustrates these data. This graph indicates that Eq. 10.9 does an excellent job of predicting the $\Delta V$ of each vehicle providing an error of several feet per second is permissible.

Equation 10.9 is not purely empirical: a rationale for its form is given in Ref. 15. In addition to its simplicity, a remarkable feature of Eq. 10.9 is that it appears to apply equally well to all of the above collision categories and conditions. With the exception of $\Delta T$, the energy loss, $\Delta V$ depends only on the mass and initial speeds of each vehicle. As noted earlier, the energy loss is relatively constant within each collision category. For collisions of the RICSAC types, experimental or least square energy loss fractions can be used as estimates. Equation 10.9 provides

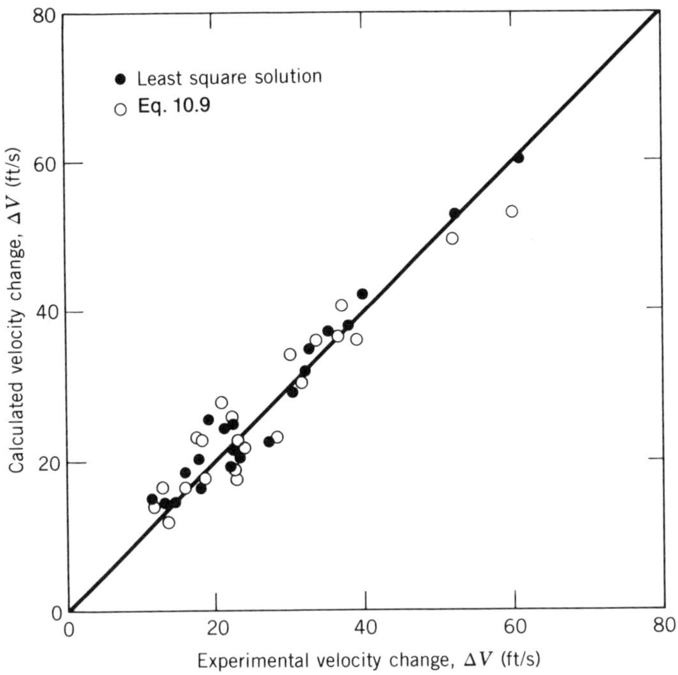


**Figure 10.4.** Comparison of least square velocity change values to experimental values.

a reasonably accurate means of estimating $\Delta V$ under these conditions. The manner in which this estimate is used depends on the nature of the problem. For socioeconomic studies of vehicle collisions, Eq. 10.9 can furnish $\Delta V$ directly as a function of initial speed if information is available concerning $\Delta T$. For accident reconstruction studies, Eq. 10.9 can be used to check and verify calculations of $\Delta V$'s.

A word of caution must be made. Although this estimate of $\Delta V$ seems to fit the 11 staged collisions rather well, Eq. 10.9 has not yet been compared to any other set of reliable collision data. Consequently, it should be used only for collisions where $e \approx 0$ and $\mu \approx \mu_0$.

## 10.5 VEHICULAR ACCIDENT RECONSTRUCTION

The analysis of vehicle collisions forms a part of a process called accident reconstruction. Accident reconstruction can be defined as the process of estimation of accident or preaccident conditions or both using postaccident information. The intention, typically, is to obtain quantitative estimates of vehicle or occupant dynamical variables. The reconstruction process generally can be broken down into three steps. These are information gathering, vehicle dynamic analyses, and evaluation and are outlined in Table 10.4. A brief discussion of the accident reconstruction process is presented subsequently and followed by an example reconstruction of a two-vehicle collision. The topic is also covered in other references.[19,20,29]

### 10.5.1 Reconstruction Techniques

The information used in the dynamic analysis step of the reconstruction typically is obtained from a variety of sources. Occasionally, important information is not available (a vehicle may have been disposed of prior to the beginning of the reconstruction) or information may have unknown reliability (such as distance and speed estimates of eyewitnesses). If vehicles are not available for measurement or inspection, photographs may have to be relied on for estimation of crush surface orientation and dimensions. Obtaining site information can be as simple as stepping off a distance between two points to more sophisticated sources such as highway department drawings, aerial photographs, and photogrammetry. In accidents with serious injuries or fatalities or both, police typically prepare postimpact diagrams with measured distances between sig-

**TABLE 10.4 Steps in a Vehicular Accident Reconstruction Process**

A. INFORMATION GATHERING
   1. Scene: observations, measurements, technical literature.
   2. Vehicles: observations, measurements, technical literature.
   3. Witnesses: statements, police reports, depositions.

B. KINEMATIC AND DYNAMIC ANALYSES
   1. Preimpact motion
   2. Impact
   3. Postimpact motion
   4. Occupant dynamics

C. EVALUATION
   1. Driver characteristics and control options
   2. Scenario comparisons
   3. Error estimation and bounding

nificant objects and a reference point. Vehicle data may come from manufacturer's specifications, industry publications, or physical measurements.

Unavailable quantitative information is often estimated, such as occupant's weight or cargo weight in vans. Missing information such as whether or not a vehicle's brakes were locked during an impact may require two analyses, one under each alternative. These comments relate to the information-gathering stage and indicate that, when significant information is missing or of dubious accuracy, the reconstruction process may have to be repeated to provide upper and lower bounds based on the uncertainties.

The next step is a dynamic analysis of the vehicles involved. A wide variety of circumstances may occur. Often the collision may be of minor significance, but the sequence of events is important. Knowing each vehicle's location and when it was there is important in establishing accident causation. Here accident kinematics is paramount. When the impact itself is significant and to be analyzed, vehicle motion is typically analyzed in three phases. These are preimpact motion, the impact motion, and postimpact motion. These three phases are usually analyzed in reverse order of occurrence. That is, the postimpact motion is analyzed first, preimpact motion last.

***10.5.1.1 Postimpact Motion.*** This analysis consists of applying a vehicle dynamics simulation individually to each vehicle's motion from the impact location to its rest position. Details of these analyses are beyond the scope of this book; see Refs. 21 and 28, for example. The kinematic conditions (velocities) are found, which, for the given environmental conditions (friction, aerodynamic drag, and so on), provide the proper rest position and orientation of each vehicle. More often than not, this is a trial-and-error procedure. For simple cases, such as a straight-ahead, locked-wheel skid, a simple hand calculation may be sufficient. Often, computerized solutions are necessary. Sometimes the motion or environment or both are too complicated or not enough information is available and an analysis is not possible.

As the result of a postimpact analysis, the initial conditions of vehicles' postimpact phase are established. These then become the final conditions of the impact phase.

***10.5.1.2 Impact Analysis.*** The reconstruction proceeds by using the appropriate solution equations (point mass, rigid body planar, and so on) with the appropriate vehicle properties and impact configuration variables. The initial impact velocities are found that match the velocities found from the postimpact analysis. Again, a trial-and-error, iterative procedure is often required. In some cases (particularly if a direct, central impact model is valid), an inverse solution of the system equations can be used and iteration is unnecessary. For more complicated collision configurations, trial-and-error iterative solutions are more common. Coefficient values from the RICSAC experiments can sometimes be used as estimates when appropriate. Upper and lower bounds on these coefficients may be necessary.

The trial-and-error process can be eliminated for the planar, rigid body model through the use of a least square approach to the solution of collision reconstruction problem,[23] but this is not discussed here.

Once the impact phase analysis is finished, the initial impact velocities are known and form the final conditions of the preimpact phase.

***10.5.1.3 Preimpact Phase.*** Often, preimpact velocities are sufficient for a complete reconstruction. Preimpact motion may be relevant, such as skidding or evasive maneuvers and may require further analysis, however. When necessary, this motion is analyzed using the same approach and analysis techniques as for postimpact motion.

As listed in Table 10.4, the final step in a reconstruction is the evaluation process. The evaluation process is highly case specific and is not discussed here in much depth. It may lead to a repetition of either or both of the first two steps. Errors in the final results such as in estimation of preimpact speeds depend on the accuracy of all of the data used in the reconstruction and can be large or small. The discussion earlier in this chapter, which compared the planar impact model solutions to experimental results, can be helpful for comparative purposes. The entire reconstruction process can rely on postimpact vehicle dynamics models and their accuracy must also be evaluated. A convenient approach is to place tolerance bounds on input data, such as tire–pavement friction coefficient or coefficient of restitution. Then the lowest or highest values are used in an appropriate combination to yield upper and lower bounds on the accident variable estimates.

## 10.5.2 Impact Analysis in Reconstruction

An outline of the reconstruction process follows with emphasis on the impact phase. Two approaches are described. The first is a direct application of the planar solution equations; the second is referred to as the CRASH method.[1] CRASH is an acronym for Calspan Reconstruction of Accident Speeds on the Highway of which CRASH3 is the current version in common use.

In some collisions, neither vehicle has significant rotation, such as a perpendicular collision of the front of one vehicle into the side of a stationary or slowly moving vehicle (see Example 2.1). Whenever rotational effects are negligible, point-mass impact theory can be used. Figure 10.5 shows a flow chart of the reconstruction process based on the planar point-mass solution equations of Chapter 2 and the corresponding concepts of limiting impulse ratio and energy loss from Chapter 3. If rest positions, point of impact, and road surface–to–tire coefficient(s) of friction are known, it may be possible to estimate reliably the velocity

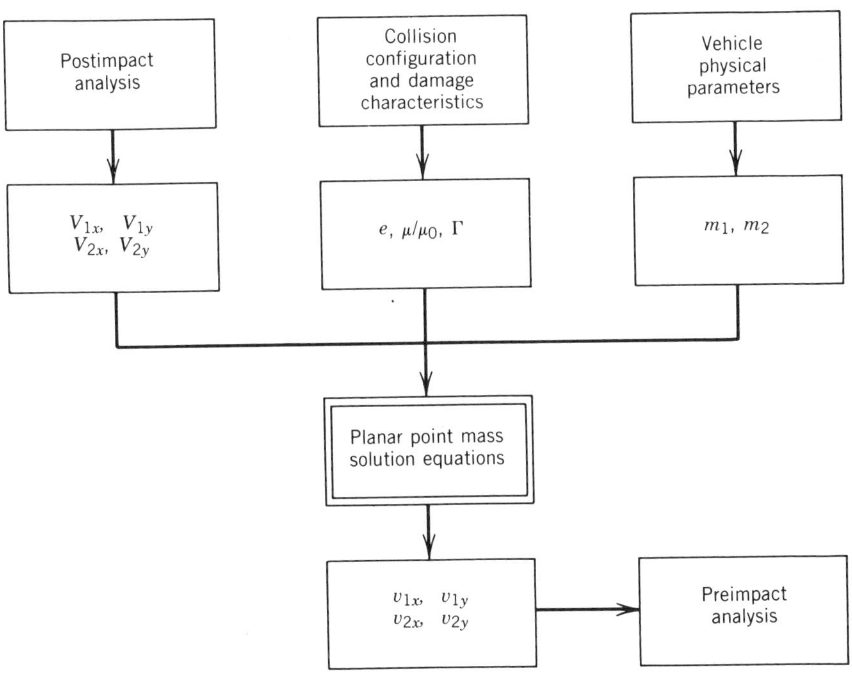


**Figure 10.5.** Accident reconstruction scheme, point-mass impact model.

of each vehicle immediately following the impact. Methods using concepts of vehicle dynamics (not covered here) must be used in a postimpact analysis to provide values of the final impact velocities. From the collision orientation (configuration) of the vehicles and the characteristics of the damage, estimates must be made of the values of $e$, $\mu/\mu_0$, and the angle $\Gamma$. The RICSAC collision data can be used as a guide. Often, "common velocity" conditions are used, which correspond to $e = 0$ and $\mu/\mu_0 = 1$.[5,6] Finally, the mass of each vehicle is needed. If $e \neq 0$, an inverse solution of the system equations can be used to furnish the initial velocities. Otherwise, the initial velocities that give the appropriate final velocities can be found by trial and error from the solution equations. Determination of the preimpact velocities may suffice for some reconstructions. For others, preimpact motion such as locked-wheel skidding, may be amenable to a further, preimpact analysis.

It should be quite apparent that the reconstruction procedure just described cannot be carried out in all cases. For example, consider a direct, head-on collision of two vehicles of equal or nearly equal mass. Suppose that in the presence of considerable damage, there is no appreciable postimpact displacement. In this case, the postimpact velocities are zero, the coefficient of restitution is zero or nearly zero, and any set of equal or nearly equal initial velocities will satisfy the solution equations. The solution of this impact reconstruction is not unique. (In this case the amount of damage and the kinetic energy loss is sensitive to the initial velocities and the CRASH method, to be discussed shortly, is applicable. Each reconstruction must be handled on an individual basis and proper judgment must be used in the selection of methods and evaluation of results.)

Solution equations for the inverse problem can be determined from Eqs. 2.12–2.15 using Eqs. 2.6 and 3.10. They are

$$v_{1n} = V_{1n} + \frac{m_2}{m_1 + m_2}\frac{1 + e}{e}(V_{2n} - V_{1n}) \qquad (10.10)$$

$$v_{1t} = V_{1t} + \frac{\mu}{\mu_0}\frac{m_2}{m_1 + m_2}\frac{V_{2t} - V_{1t}}{1 - \mu/\mu_0} \qquad (10.11)$$

$$v_{2n} = V_{2n} - \frac{m_1}{m_1 + m_2}\frac{1 + e}{e}(V_{2n} - V_{1n}) \qquad (10.12)$$

and

$$v_{2t} = V_{2t} - \frac{\mu}{\mu_0} \frac{m_1}{m_1 + m_2} \frac{V_{2t} - V_{1t}}{1 - \mu/\mu_0} \tag{10.13}$$

Note that solutions do not exist for $e = 0$ and $\mu/\mu_0 = 1$. In some cases the magnitude of the velocity change of the vehicles is of interest. These are found from the magnitude of the momentum change, which is

$$m\Delta V_i = m_i\left[(V_{in} - v_{in})^2 + (V_{it} - v_{it})^2\right]^{1/2} \tag{10.14}$$

Using the solution equations of Chapter 2 and the expression for the energy loss, Eq. 3.5, this can be written as

$$m_i \Delta V_i = \left[ \frac{2\overline{m}(1 + e)^2(1 + \mu^2)T_L}{(1 - e^2) + r^2 \dfrac{\mu}{\mu_0}\left(2 - \dfrac{\mu}{\mu_0}\right)} \right]^{1/2}, \qquad i = 1, 2 \tag{10.15}$$

For the common velocity conditions, $e = 0$ and $\mu = \mu_0$,

$$m_i \Delta V_i = (2\overline{m}T_L)^{1/2}, \qquad i = 1, 2 \tag{10.16}$$

Knowledge of the energy loss in this case leads directly to the velocity change magnitude of each vehicle.

Figure 10.6 shows a flow chart corresponding to the popular CRASH method (under point-mass conditions). Two options are available and are indicated by solid and dashed blocks on the diagram. The solid blocks illustrate an approach comparable to the one just covered. CRASH has a built-in method of post impact analysis.

As a result of its formulation and assumptions, CRASH is limited to the common velocity conditions and also requires that the direction of impulse (DOI)* be specified as an input quantity. The second option (dashed lines) is one that does not use a postimpact analysis of the vehicle dynamics, but is based on measurement of the amount of defor-

*Actually the CRASH programs use what is called a PDOF, principal direction of force. This quantity is undefined in the CRASH literature, but seems to correspond to the line of action of the resultant impulse.

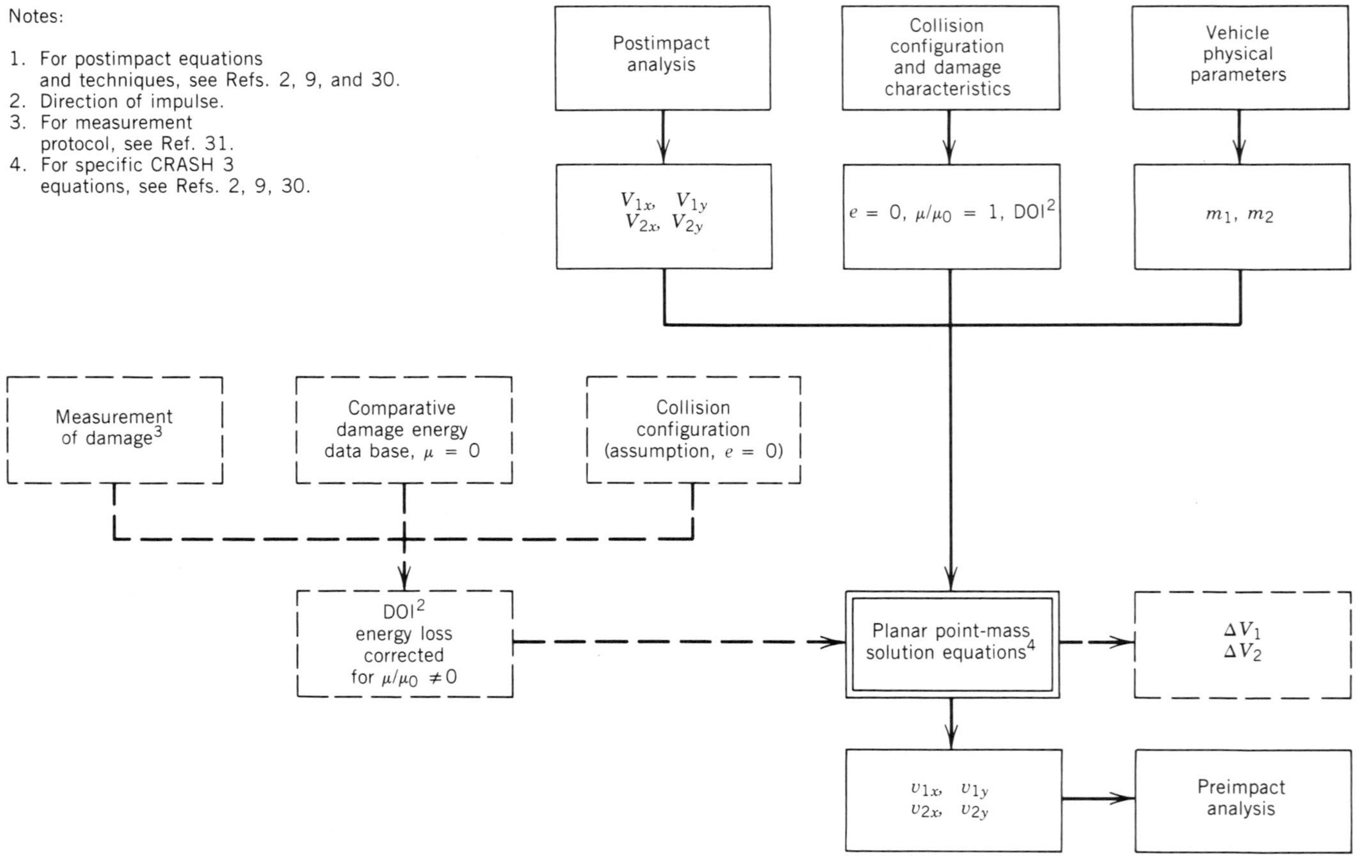


**Figure 10.6.** CRASH3 accident reconstruction scheme, point-mass impact model: solid blocks indicate damage and spinout; dashed blocks indicate damage only, no postimpact analysis.

mation or crush of each vehicle. Comparison of the extent and depth of crush of the actual vehicles is made to an experimentally determined crush data base that provides an estimate of the kinetic energy loss for the collision. Use of Eq. 10.16 then provides the velocity change magnitudes for the vehicles. With the direction of impulse specified, the $\Delta V$'s are fully known and can be used, for example, to analyze occupant dynamics.

When rotational effects are not negligible or are otherwise of interest, the reconstruction must be based on the rigid body solution equations. Figure 10.7 illustrates the corresponding scheme. In principle, it does not differ from the point-mass approach. The planar equations of Chapter 5 are used along with the related expressions for energy loss and the critical impulse ratio. Computer solutions are usually the only practical way of solving the equations particularly since trial-and-error solutions are often needed. The inverse solution is also best done numerically and so the rigid body counterparts of Eq. 10.10–10.13 are not presented. It

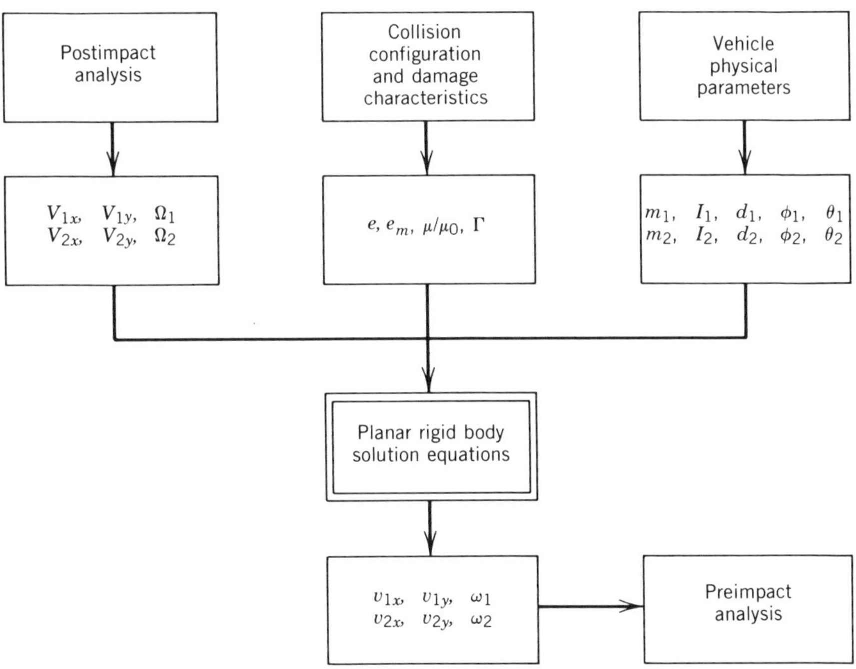


**Figure 10.7.** Accident reconstruction scheme, planar rigid body model.

is possible to determine an expression for the velocity change magnitude (for $e_m = -1$), which is

$$m_i \Delta V_i = \left[ \frac{2\overline{m}(1 + e)^2(1 + \mu^2) q T_L}{(1 - e^2) + r^2 \frac{\mu}{\mu_c}\left(2 - \frac{\mu}{\mu_c}\right)} \right]^{1/2}, \quad i = 1, 2 \qquad (10.17)$$

where $q$ is given by Eq. 5.33. Not enough experimental work has been done to fully determine the range of practical values of $\mu_c$ for automobile collisions. It appears,[13] however, that in practice, since $e$ is usually small, the limiting impulse ratio for vehicle collisions is $\mu_c = \mu_0$. For $\mu = \mu_c = \mu_0$ and $e = 0$, Eq. 10.17 gives

$$m_i \Delta V_i = [2\overline{m} q T_L]^{1/2} \qquad (10.18)$$

for those collisions, which satisfy the common velocity condition at the time of separation.

The approach of Fig. 10.6 still applies to the rigid body formulation of CRASH3 provided that the appropriate rotational velocities, vehicle dimensions, and vehicle inertial properties are now included. Users of the CRASH method should become familiar with the simplifying assumptions of that method, some of which are unstated. For example, an approximate version of Eq. 10.17 is used by CRASH3 to compute $\Delta V_i$. For the damage-measurement alternative, questions can be raised about the basis upon which the direction of impulse (DOI) is established and how the energy loss from the data base is corrected[9] for the condition of $\mu = \mu_0$. Also the CRASH programs will not calculate a momentum solution for head-on or near head-on collisions (when the initial velocity vectors differ in orientation by less than 10°). This limitation is a consequence of an unfortunate choice of coordinate systems.

### 10.5.3 Sample Reconstruction

A rather comprehensive example is now presented of a reconstruction of an actual collision. A postimpact analysis is not presented; results from such an analysis for each of the two vehicles simply are provided.

## ■ *Example 10.1*

PROBLEM: Figure E10.1.1 illustrates a scale drawing of the scene of a nighttime accident. It shows the final rest positions of the two vehicles. Figure E10.1.2 shows the impact positions and orientations as determined from vehicle damage patterns, pavement markings, and debris location (the indicated roadside debris was primarily rear window glass). Rest position coordinates were obtained from measurements indicated on a police accident diagram; utility poles 1 and 2 were used as reference points for scene measurements. Vehicle 1 is a two-door sedan and had four occupants; vehicle 2 is an unladened pickup truck with a driver only.

A postimpact analysis of Veh 2 indicates a speed of about 55 ft/s (37.5 mph) after the impact with an orientation of 0° to 10° to the road

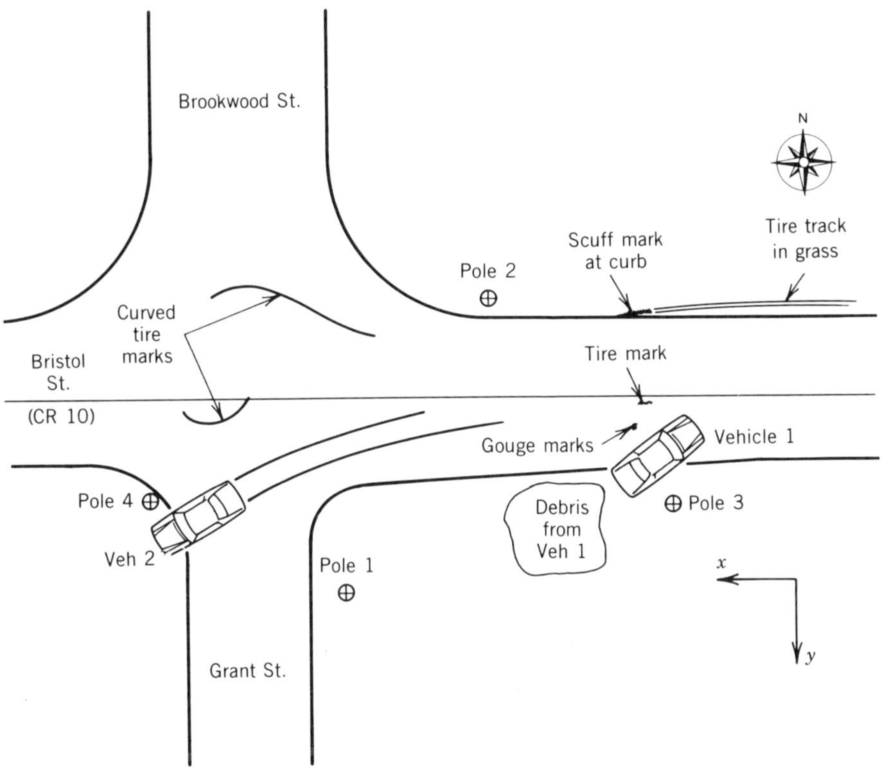


**Figure E10.1.1.** Accident scene.

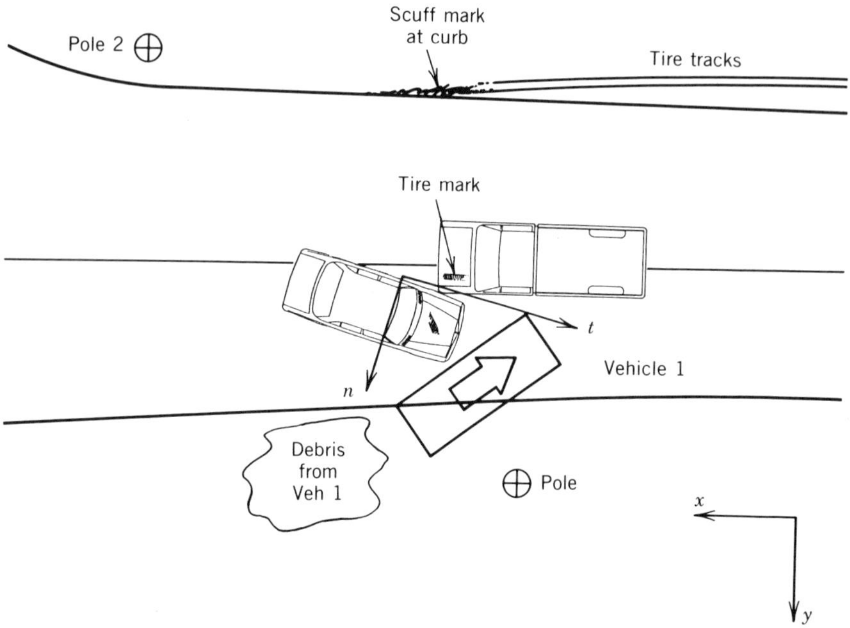


**Figure E10.1.2.** Impact configuration.

centerline. Its tire skid marks indicate little or no postimpact angular velocity. A postimpact analysis of Veh 1 shows a corresponding velocity of about 8 ft/s (5.5 mph) in a southeasterly direction, oriented about 75° from the road centerline and with an angular velocity between 1.5 and 2.0 rad/sec. See Fig. E10.1.2 for positions and orientations.

Use the planar impact model to estimate the initial, or preimpact, speeds of both vehicles. Figure E10.1.3 shows the important vehicle parameters that enter into the impact analysis.

*Solution:* No evidence indicates a significant moment impulse, so $e_m$ is chosen to equal $-1$. Damage was quite extensive to both vehicles, so $e$ was chosen near zero at $e = 0.05$. The angle $\Gamma$ was chosen such that the crush surface is parallel to the side of Veh 1. Experimental evidence indicates that $\mu \sim \mu_0$ for many collisions. One exception, however, is the sideswipe collision of which this is an example. The impulse ratio was simply estimated to be 1.5 with a negative sense; thus, $\mu = -1.5$. (More discussion of this later.)

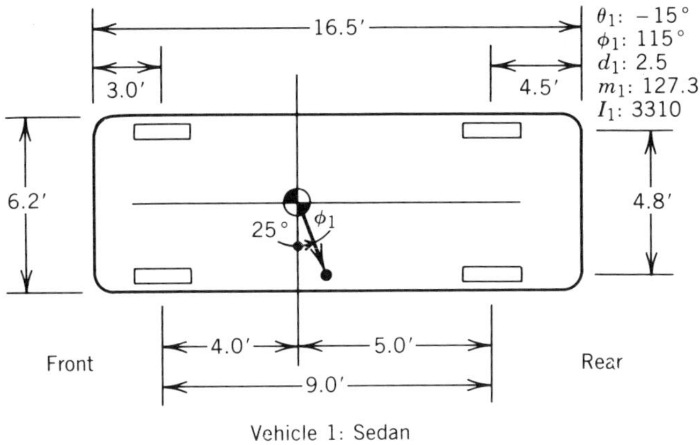


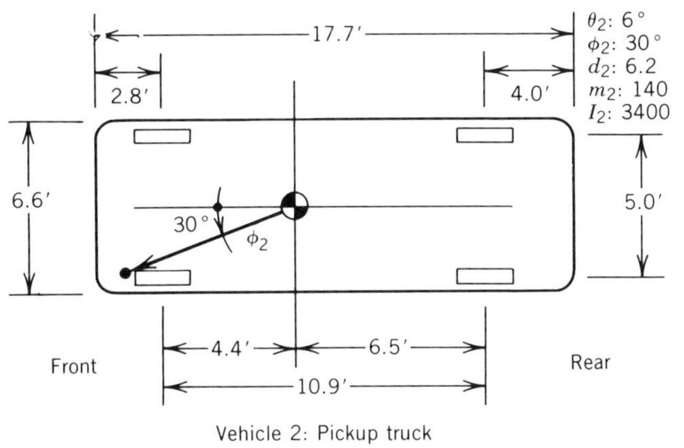


**Figure E10.1.3.** Physical data for vehicles.

A trial-and-error selection of initial velocity components for Veh 1 and Veh 2 using a computerized solution of the system equations produced the results shown in Table 10.1.1. According to this solution, Veh 2 had an initial velocity of 70.0 ft/s (48 mph), directly west. Veh 1 had a velocity of 15.1 ft/s (10.3 mph) at an angle of about 8° to the road centerline, east by southeast. Both had small initial angular velocities. Almost 40% of the system kinetic energy was lost, about 36% due to tangential effects.

A second solution was found using a technique that seeks an inverse solution using the method of least squares.[23] In that solution scheme,

**TABLE E10.1.1 Impact Analysis of a Field Example**
**Estimates of Velocity Changes (US Units, ft, lb, s)**

Restitution coefficient, $e = 0.050$ Moment coefficient, $e_m = -1.00$
Impulse ratio, $\mu = -1.500$ Crush angle, $\Gamma = 75.0°$
Impulse ratio, $\mu_0 = -2.440$

| Vehicle 1 | Vehicle 2 |
|---|---|
| $m = 127.3$ $I = 3310.0$ | $m = 140.0$ $I = 3400.0$ |
| $\phi = 115.0$ $\theta = -15.0$ | $\phi = 20.0$ $\theta = 0.0$ |
| $d_1 = 2.50$ | $d_2 = 6.50$ |

PHYSICAL PARAMETERS

| Vehicle 1 | Vehicle 2 |
|---|---|
| $v(x) = -15.00$ $v(n) = -1.95$ | $v(x) = 70.00$ $v(n) = 18.12$ |
| $v(y) = 2.00$ $v(t) = 15.01$ | $v(y) = 0.00$ $v(t) = -67.61$ |
| $\omega = -0.05$ | $\omega = -0.10$ |

INITIAL VELOCITIES

| Vehicle 1 | Vehicle 2 |
|---|---|
| $V(x) = 2.24$ $V(n) = 8.15$ | $V(x) = 54.32$ $V(n) = 8.94$ |
| $V(y) = 7.83$ $V(t) = -0.14$ | $V(y) = -5.30$ $V(t) = -53.84$ |
| $\Omega = 1.68$ | $\Omega = 0.00$ |

FINAL VELOCITIES

$\Delta V_1 = 18.20$ $\Delta V_2 = 16.55$

| Initial | Final | | Initial | Final |
|---|---|---|---|---|
| 0.1458E+05 | 0.4226E+04 | ⟨translational⟩ | 0.3430E+06 | 0.2085E+06 |
| 0.4138E+01 | 0.4672E+04 | ⟨rotational⟩ | 0.1700E+02 | 0.2780E−02 |
| 0.1458E+05 | 0.8898E+04 | ⟨total⟩ | 0.3430E+06 | 0.2085E+06 |

KINETIC ENERGY

Initial system energy = 3.5760E+05 Crush energy loss = 3.3%
Maximum energy loss = 56.9% Frictional energy loss = 35.9%
Moment crush energy loss = 0.0% System energy loss = 39.2%

Impulse, $P(x) = 2195.1$ Normal, $P(n) = 1285.4$
Impulse, $P(y) = 742.6$ Tangential, $P(t) = -1928.1$
Resultant impulse, $P = 2317.3$ Moment impulse, $M = 0.0$

specified final impact velocity component values are matched in a least squares sense. In the process, any combination of up to six unknowns are calculated (limited by the number of equations) where the unknowns can be initial velocity components or coefficients or both. Table 10.1.2 shows the results of this process. The moment coefficient was again cho-

**TABLE E10.1.2 Impact Analysis of a Field Example**
**Least Square Estimates of Velocity Changes (US Units, ft, lb, s)**

| | |
|---|---|
| Restitution coefficient, $e = 0.050$ | Moment coefficient, $e_m = -1.00$ |
| Impulse ratio, $\mu = -0.554$ | Crush angle, $\Gamma = 75.0°$ |
| Impulse ratio, $\mu_0 = -2.038$ | |

| Vehicle 1 | | Vehicle 2 | |
|---|---|---|---|
| $m = 127.3$ | $I = 3310.0$ | $m = 140.0$ | $I = 3400.0$ |
| $\phi = 115.0$ | $\theta = -15.0$ | $\phi = 20.0$ | $\theta = 0.0$ |
| $d_1 = 2.50$ | | | $d_2 = 6.50$ |

PHYSICAL PARAMETERS

| | | | |
|---|---|---|---|
| $v(x) = -3.85$ | $v(n) = 0.00$ | $v(x) = 59.86$ | $v(n) = 15.49$ |
| $v(y) = 1.03$ | $v(t) = 3.99$ | $v(y) = 0.00$ | $v(t) = -57.82$ |
| $\omega = 0.94$ | | | $\omega = 0.89$ |

INITIAL VELOCITIES

| | | | |
|---|---|---|---|
| $V(x) = 2.20$ | $V(n) = 7.61$ | $V(x) = 54.36$ | $V(n) = 8.57$ |
| $V(y) = 7.29$ | $V(t) = -0.23$ | $V(y) = -5.70$ | $V(t) = -53.98$ |
| $\Omega = 1.62$ | | | $\Omega = -0.04$ |

FINAL VELOCITIES

$\Delta V_1 = 8.71$ $\Delta V_2 = 7.92$

| Initial | Final | | Initial | Final |
|---|---|---|---|---|
| 0.1011E+04 | 0.3693E+04 | ⟨translational⟩ | 0.2508E+06 | 0.2091E+06 |
| 0.1462E+04 | 0.4328E+04 | ⟨rotational⟩ | 0.1347E+04 | 0.2613E+01 |
| 0.2473E+04 | 0.8021E+04 | ⟨total⟩ | 0.2522E+06 | 0.2091E+06 |

KINETIC ENERGY

| | |
|---|---|
| Initial system energy = 2.5465E+05 | Crush energy loss = 3.5% |
| Maximum energy loss = 39.9% | Frictional energy loss = 11.2% |
| Moment crush energy loss = 0.0% | System energy loss = 14.7% |
| Impulse, $P(x) = 769.6$ | Normal, $P(n) = 969.4$ |
| Impulse, $P(y) = 797.4$ | Tangential, $P(t) = -537.0$ |
| Resultant impulse, $P = 1108.2$ | Moment impulse, $M = 0.0$ |

sen as $e_m = -1$. The other coefficients were not chosen, but were estimated by the least square fitting process to be $e = 0.050$ and $\mu = -0.554$. The earlier estimate of $e$ was returned exactly, but the magnitude of the impulse ratio, 0.554, is now much smaller than the 1.5 used above. Other changes occur with the initial velocities now becoming

3.99 ft/s (2.7 mph) and 59.9 ft/s (41 mph) for Veh 1 and Veh 2, respectively.

Solutions for both values of $\mu$ indicate an unusually low preimpact speed for Veh 1, of 4–10 mph. This low speed along with other evidence (tire track in the north berm grass and curved tire marks in the intersection) support a hypothesis that Veh 1 had just completed an erratic U-turn prior to the collision. This in turn indicates that Veh 2 may have been to the left of center as a maneuver to avoid Veh 1 and contradicts a hypothesis that the accident was caused by Veh 2 drifting into the opposing lane of traffic. ■

## 10.6 IMPACT OF LIGHT AIRCRAFT WITH THE GROUND

The specific reasons for analyzing the collision of an aircraft either with a structure or the ground often differ. Typically, all have safety as an ultimate objective. One may be interested in the response of the structure, such as for the safety of nuclear power plants,[25] whereas another may be interested in the response of the aircraft structure and interior[8] with concern for occupant safety. In these cases, dynamical plastic analyses are often necessary to establish the force–deformation–stress time histories of the pertinent structures in order to quantify damage, failure, and accelerations. If the collision dynamics of the aircraft itself are of interest and the aircraft does not break up, the approach of this book can often furnish reasonably accurate estimates. If triangular force–time histories are appropriate, accelerations can be estimated. Most studies associated with nuclear containment assume a worst-case scenario, where the impact is at high speed, perfectly inelastic, and normal to the structure[26]; little or no experimental information is available from that field for oblique impact. Some experimental aircraft-to-ground collisions have been conducted by NASA.[8] These collisions were conducted primarily to determine cockpit and other interior accelerations for structural analysis and occupant safety and survival purposes. Rigid body dynamic data such as rotational velocities of the plane's axes during or following impact were not completely determined. Consequently, the analysis of these data, which follows, should be considered incomplete. Some results remain open to question, but this is a first step in this area, so the results are presented and discussed to stimulate further work.

All of the data to be discussed correspond to relatively light weight, general aviation aircraft as tested in controlled crashes. Of the many tests conducted, only nine have a complete enough set of published data suitable for analysis here. The aircraft were all swung by cables in a vertical plane with various heading or pitch angles and flight path angles from an overhead structure to impact a flat surface as shown in Fig. 10.8. Side-to-side symmetry was reasonably well maintained, which permits a planar analysis. The following comments and assumptions can be made:

- Published photographs show little vertical bounce (likely due to the nature of the structural crush) and a coefficient of restitution, $e = 0$, is used for all crashes.
- After the impact, each plane continued to slide for a considerable distance along the ground. Analysis of these data[8] shows a fairly consistent friction coefficient of $f = 0.4$. This value is used here in all crashes.
- The impact duration was determined by the signal from transducers mounted both perpendicular to and along the longitudinal axis of the airframe. In some cases these gave different values of impact

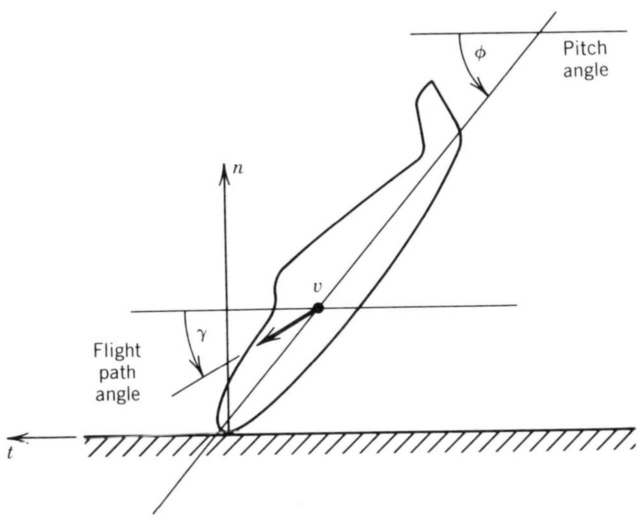


**Figure 10.8.** Aircraft crash orientation.

duration. The larger of the two, when they differ, is used here arbitrarily.

- Upon impact, the rear fuselage in some cases slapped the ground. Others, owing to extended rigid landing gear, remained oriented closer to their precrash tail-up heading. Individual conditions are not indicated in published data.

Neglecting the effects in the last item amounts to an assumption that the moment impulse is negligible, that is, $M = 0$ and $e_m = -1$. This may be inappropriate, but must be done here for lack of information.

With these assumptions, the calculations are quite simple. With $e = 0$, the mass and initial vertical velocity provide the normal impulse $P_n$. That is

$$P_n = mv_n = mv \sin \gamma \tag{10.19}$$

where $v$ is the initial speed and $\gamma$ is the flight path angle. The measured slideout distance $d$ and coefficient of friction allow the final tangential velocity $V_t$ to be estimated. The tangential impulse is

$$P_t = m(V_t - v_t) = m(\sqrt{2fgd} - v \cos \gamma) \tag{10.20}$$

Table 10.5 shows the experimental data and Table 10.6 lists the calculated quantities.

Analysis of the data by NASA personnel has lead to the conclusion that of triangular, rectangular, and half-sine curves, the triangular shape does the best job of representing the experimental pulse shapes. As shown in Chapter 2, knowledge of the pulse shape allows calculation of peak forces (and accelerations). The peak or maximum accelerations normal and tangential to the ground and based on triangular pulse shapes are listed in Table 10.6. Acceleration components were measured perpendicular and along the roll axis of the planes. The orientation of these transducers were changing throughout the impact, and so components cannot be compared or even transformed to enhance comparison. Comparison of the peak (maximum) total accelerations can be made but it must be kept in mind that the measured peak values along each axis may not have occurred at the same instant. Peak total accelerations are listed in Tables 10.5 and 10.6. The calculated values differ in magnitude by

**TABLE 10.5 Experimental Light Aircraft Crash Data[8]**

| NASA Test Number | Flight Path Angle (deg) | Fuselage Pitch Angle (deg) | Flight Path Velocity (m/s) | Total Mass (kg) | Measured Slideout Distance (m) | Impact Duration (s) | Measured Peak Acceleration (g's) |
|---|---|---|---|---|---|---|---|
| 2 | 16.0 | 12.0 | 27.0 | 2671 | 48.0 | 0.089 | 27.586 |
| 3 | 18.8 | 18.0 | 26.2 | 2751 | 54.0 | 0.050 | 33.287 |
| 6 | 16.0 | −14.0 | 27.0 | 2700 | 59.0 | 0.110 | 19.698 |
| 7 | 47.5 | 47.3 | 28.6 | 2700 | 26.5 | 0.174 | 21.850 |
| 8 | 30.0 | 31.0 | 27.0 | 2700 | 37.5 | 0.135 | 24.083 |
| 13 | 29.0 | 26.0 | 25.0 | 1090 | 47.0 | 0.093 | 29.155 |
| 18 | 30.0 | 31.0 | 27.9 | 1993 | 56.0 | 0.090 | 31.159 |
| 19 | 15.0 | 17.7 | 27.0 | 2193 | 78.0 | 0.120 | 16.919 |
| 20 | 15.4 | −2.0 | 26.6 | 2239 | 77.0 | 0.057 | 31.654 |

4.0% to 28.6% from their measured counterparts, which is illustrated in Fig. 10.9.

Perhaps the most unexpected result of the calculations is the wide range of experimental values of $\mu$, the impulse ratio, from a low of 0.150 to a high of 0.879. Other than initial speeds and angles only two observed quantities enter into the calculation of $\mu$. These are the slideout

**TABLE 10.6 Analysis of Aircraft Crash Data[a]**

| NASA Test Number | Slideout Momentum (kg-m/s) | Normal Impulse (N-s) | Tan'l Impulse (N-s) | Impulse Ratio $\mu$ | $\ddot{N}$,max (g's) | $\ddot{T}$,max (g's) | Triangular Pulse Peak (g's) |
|---|---|---|---|---|---|---|---|
| 2 | 51,841 | 19,878 | 17,482 | .879 | 17.048 | 14.993 | 22.703 |
| 3 | 56,633 | 23,168 | 11,618 | .501 | 34.339 | 17.221 | 38.415 |
| 6 | 58,099 | 20,094 | 11,977 | .596 | 13.793 | 8.221 | 16.058 |
| 7 | 38,937 | 56,933 | 13,232 | .232 | 24.706 | 5.742 | 25.365 |
| 8 | 46,319 | 36,450 | 16,814 | .461 | 20.387 | 9.405 | 22.452 |
| 13 | 20,934 | 13,211 | 2,899 | .219 | 26.570 | 5.831 | 27.202 |
| 18 | 41,781 | 27,802 | 6,374 | .229 | 31.600 | 7.245 | 32.420 |
| 19 | 54,258 | 15,325 | 2,935 | .192 | 11.872 | 2.274 | 12.088 |
| 20 | 55,040 | 15,816 | 2,379 | .150 | 25.265 | 3.800 | 25.549 |

[a]In all cases the coefficient of restitution $e$ is zero and the coefficient of friction for slideout is 0.4.

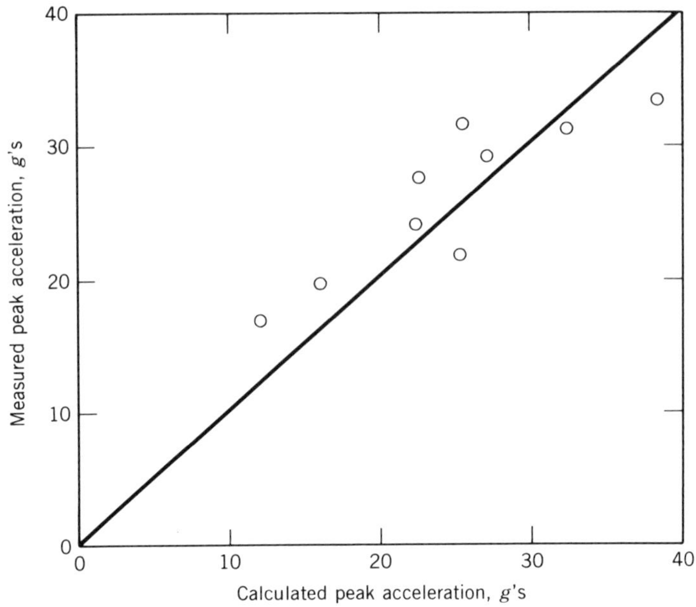


**Figure 10.9.** Maximum accelerations by measurement and triangular pulse approximation.

distance and the coefficient of friction. NASA personnel carried out a specific analysis[8] of the friction coefficient in the slideout phase and present the value of $f = 0.4$ as consistent and reliable. One would expect the measured slideout distances to have negligible error. All of this leads to speculation for the reason(s) for the wide range of $\mu$ values and suggests that additional tests need to be conducted or additional data from existing tests should be analyzed or both.

## REFERENCES


1. McHenry, R. R., "Extensions and Refinements of the CRASH Computer Program, Part I, Analytical Reconstruction of Highway Accidents," DOT-HS-5-01124, Feb., 1976.
2. Anon., *CRASH 3 Users Guide and Technical Manual*, U.S. DOT, NHTSA, Washington, D.C. 29590.
3. Ishekawa, H., "Computer Simulation of Automobile Collisions—Reconstructions of Accidents," Paper 851729, SAE, Warrendale, PA 15096.

4. Macmillan, R. H., *Dynamics of Vehicle Collisions*, Interscience, New York, 1983.
5. Wooley, R. L., ''The 'IMPAC' Computer Program for Accident Reconstruction,'' Paper No. 850254, SAE, Warrendale, PA 15096.
6. Brach, Raymond M. and Brach, R. Matthew, ''A Review of Impact Models for Vehicle Collision,'' Paper No. 870048, SAE, Warrendale, PA 15096.
7. Brach, R. M., ''Impact of Articulated Vehicles,'' Paper No. 860015, SAE, Warrendale, PA 15096.
8. Carden, H. D., ''Impact Analysis of Airplane Crash Data with Consideration Given to Human Tolerance,'' Paper No. 830748, SAE, Warrendale, PA 15096.
9. Wooley, R. L., Warner, C. Y., and Tagg, M. D., ''Inaccuracies in the CRASH3 Program,'' in *Field Accidents: Data Collection, Analysis, Methodologies, and Crash Injury Reconstruction*, SAE P-159, Warrendale, PA, 1985.
10. Brach, R. M., ''Energy Loss in Vehicle Collisions,'' Paper No. 871993, SAE, Warrendale, PA 15096.
11. Jones, I. S. and Baum, A. S., *Research Input for Computer Simulation of Automobile Collisions, Vol. IV, Staged Collision Reconstructions*, DOT HS-805040, Dec., 1978.
12. Smith, R. A. and Noga, J. T., ''Accuracy and Sensitivity of CRASH,'' Paper No. 821169, SAE, Warrendale, PA 15096.
13. Brach, R. M., ''Impact Analysis of Two-Vehicle Collisions,'' Paper No. 830468, SAE, Warrendale, PA 15096.
14. Brach, R. M., ''Nonlinear Parameter Estimation of a Vehicle Collision Model,'' Proc. of 13th Ann. Pittsburgh Conf. on Modeling and Simulation, U.P.H., 1982, pp. 525–532.
15. Brach, R. M., ''Momentum and Energy Analysis of Automobile Collisions,'' *Proc. Structural Impact and Crashworthiness*, Elsevier, 1984.
16. Emori, R. I., ''Vehicle Mechanics of Intersection Collision Impact,'' Paper No. 700177, SAE, Warrendale, PA 15096.
17. Grime, G. and Jones, J. S., ''Car Collisions—The Movement of Cars and Their Occupants in Accidents,'' Proc. Instn. Mech. Engrs., Vol. 184, Pt. 2A, No. 5, 1969–70.
18. Moore, D. F., ''Theoretical Prediction of the Trajectory of Automobiles After Side-Impact,'' Proc. 1st Int. Conf. Veh. Mech., Wayne St. Univ., July 1968.
19. Baker, J. S. and Fricke, L. B., *The Traffic-Accident Investigation Manual*, Northwestern Univ. Traffic Inst., 1986.

20. Grime, G., *Handbook of Road Safety Research*, Butterworths, London, 1987.
21. Ellis, J. R., *Vehicle Dynamics*, Business Books, Ltd., London, 1969.
22. Bundorf, R. T., "Directional Control Dynamics of Automobile-Travel Trailer Combinations," SAE 670099, 1967.
23. Brach, R. M., "Least Squares Collision Reconstruction," Paper No. 870429, SAE, Warrendale, PA 15096.
24. Riera, J. D., "A Critical Reappraisal of Nuclear Power Plant Safety Against Accidental Aircraft Impact," *Nuclear Engineering and Design*, **57**, 193–206, 1980.
25. Gardner, J. W., "Calculation of the Forces Acting Upon a Rigid Structure from an Aircraft Impact," *Int. J. Impact Engrg.*, **2**(4), 345–356, 1984.
26. Fisher, A., "Propelled Jet, Science Newsfront," *Popular Science Magazine*, December 1989, pp. 14, 16.
27. Prasad, A. K., "Energy Dissipated in Vehicle Crush—A Study Using the Repeated Test Technique," SAE Technical Paper No. 900412, SAE Congress, Detroit, 1990.
28. Wong, J. W., *Theory of Ground Vehicles*, Wiley-Interscience, New York, 1978.
29. Backaitis, S. H. (ed.), *Reconstruction of Motor Vehicle Accidents: A Technical Compendium*, PT-34, SAE, Warrendale, PA 15096.
30. Day, T. D. and Hargens, R. L., "An Overview of the Way EDCRASH Computes Delta-V," Paper No. 870045, SAE, Warrendale, PA 15096.
31. Tumbas, N. S. and Smith, R. A., "Measurement Protocol for Quantifying Vehicle Damage from an Energy Basis Point of View," in *Field Accidents: Data Collection, Analysis, Methodologies and Crash Injury Reconstruction*, SAE P-159, Warrendale, PA 15096, 1989.

# INDEX